Natural Resistance of Plants to Pests

ACS SYMPOSIUM SERIES **296**

Natural Resistance of Plants to Pests

Roles of Allelochemicals

Maurice B. Green, EDITOR
Independent Consultant

Paul A. Hedin, EDITOR
Boll Weevil Research Laboratory

Developed from a symposium sponsored by
the Division of Pesticide Chemistry
at the 189th Meeting
of the American Chemical Society,
Miami Beach, Florida,
April 28–May 3, 1985

American Chemical Society, Washington, DC 1986

Library of Congress Cataloging in Publication Data

Natural resistance of plants to pests.
(ACS symposium series, ISSN 0097-6156; 296)

"Developed from a symposium sponsored by the
Division of Pesticide Chemistry at the 189th Meeting of
the American Chemical Society, Miami Beach, Florida,
April 28–May, 1985."

Includes indexes.

1. Allelopathic agents—Congresses. 2. Pests—
Biological control—Congresses. 3. Insect–plant
relationships—Congresses.

I. Green, Maurice B. (Maurice Berkeley) II. Hedin,
Paul A. (Paul Arthur), 1926–   . III. American
Chemical Society. Division of Pesticide Chemistry.
IV. American Chemical Society. Meeting (189th: 1985:
Miami Beach, Fla.) V. Series.

QK898.A43N38   1986   581.5′249   85–28748
ISBN 0–8412–0950–2

FOREWORD

The ACS SYMPOSIUM SERIES was founded in 1974 to provide a medium for publishing symposia quickly in book form. The format of the Series parallels that of the continuing ADVANCES IN CHEMISTRY SERIES except that, in order to save time, the papers are not typeset but are reproduced as they are submitted by the authors in camera-ready form. Papers are reviewed under the supervision of the Editors with the assistance of the Series Advisory Board and are selected to maintain the integrity of the symposia; however, verbatim reproductions of previously published papers are not accepted. Both reviews and reports of research are acceptable, because symposia may embrace both types of presentation.

CONTENTS

vii

ALLELOCHEMICALS AS PEST CONTROL AGENTS

INDEXES

PREFACE

INSECT PESTS AND PATHOGENIC FUNGI cause enormous losses of food and fiber crops throughout the world. These losses tend to obscure the fact that most plants are resistant to most pests and diseases, and that attack occurs only when, as a result of coevolution, the pest or parasite has been successful in overcoming the natural defenses of the host. These defenses may be morphological. For example, the leaf surface may not be conducive to feeding or oviposition by the insect or to spore germination or mycelial growth of the fungus. But, in many cases, the defense mechanism involves the production of endogenous chemicals that deter feeding or oviposition or inhibit spore germination or mycelial growth.

Until fairly recently, little was known about the chemical structures of these endogenous substances. Professor Louis Wain, the keynote speaker of the symposium upon which this book is based, pioneered work in this field at Wye College in England forty years ago. In the last few years, a surge of investigations worldwide has resulted in the elucidation of the chemical structures of many of the endogenous substances that protect plants against attack by fungi—the phytoalexins—and in some understanding of the biochemistry of their modes of action. Much work has been done to isolate and characterize the endogenous substances that deter insects. It was, therefore, timely to organize this symposium to review the information obtained so far and to consider current research findings and prospects for the future.

The picture that emerged was a bewildering array of diverse chemicals with little, if any, common connection. So far, none of the natural endogenous chemicals has proved to be useful for crop protection in practical agriculture because many of the structures are too complex to permit economical manufacture, and partly because the substances are produced in nature at or near the site of action, and therefore are not effective when applied topically. However, a knowledge of the chemical structures may serve as a guide for the synthesis of analogous compounds that may be, under appropriate conditions, even more effective than the natural substances and more economical to manufacture. The feasibility of this approach is substantiated by the great success that has been achieved in producing useful agrochemicals based on the structure of the natural insecticide, pyrethrum, but with much greater utility, higher activity, and lower cost than the natural substance.

ix

Furthermore, an understanding of the biochemistry, and particularly the genetic mechanisms of the natural resistance of plants to pests and diseases, may open up the possibility of transferring this resistance from resistant to susceptible species. This field has a very exciting future, some idea of which can be envisaged from the chapters in this volume.

MAURICE B. GREEN
23 Brunswick Road
Withington, Manchester M2O 9QB
England

PAUL A. HEDIN
Boll Weevil Research Laboratory
Mississippi State, MS 39762

OVERVIEW

1

Developing Research Trends in the Chemistry of Plant Resistance to Pests

Paul A. Hedin

Boll Weevil Research Laboratory, USDA, Agricultural Research Service, Mississippi State, MS 39762

The development of crop plant varieties with
resistance to insects and diseases has several
advantages compared with other approaches to pest
control because of economics, environmental considerations, and relative efficiency. In recent
years, chemistry has begun to have an impact on
selection of resistant lines. Growing amounts of
information have been synthesized by chemists and
cooperating biologists on the presence and roles of
allelochemicals in resistant plants, suggesting that
its application to selections may now be a realistic
strategy. Developing research trends in the chemistry of plant resistance to pests are identified
and reviewed.

The development of crop plant varieties with resistance to pests
including insects and diseases has several advantages compared
with other approaches to pest control. The application of pesticides may be decreased or eliminated, resulting in decreased costs
of production and environmental pollution. The use of resistant
varieties also does not require the adoption of complicated tech-
nologies by the user.

Until now, resistant varieties have been selected from field
evaluations by plant breeders, often in cooperation with entomologists, plant pathologists, and/or other biologists. In recent
years, chemists have identified a number of compounds conferring
resistance that had previously or subsequently been shown to be
controlled by single genes or combinations of genes.
Investigations in which a relatively large number of available
susceptible and resistant lines or varieties have been analyzed to
determine whether correlations between the allelochemical and
resistance are statistically significant have only recently been
carried out on a substantive scale. It is therefore generally
accurate to state that the impact of chemistry (and chemists) on
the selection of resistant lines by plant breeders has been mini-
mal. On the other hand, the growing amount of information that

has recently been generated by chemists and cooperating biologists
on the presence and roles of chemicals in resistant plants
suggests that its application to plant breeding may now be
realistic. Efforts to define the roles of these chemicals in
various plant-insect and plant-plant interactions governing plant
resistance have been reported in a succession of conferences, symposia, and their proceedings, especially during the last 10 years
(1-4).
    This chapter is an effort to identify the various concepts
and approaches currently being used by scientists to chemically
define the parameters of plant resistance to pests. Of the thir-
teen approaches that were identified, only three, each associated
with behavior, were unique to insects. The remaining 10 are
generally applicable to either insects or diseases, and were
therefore not separated. The listing of examples of these may
suggest emerging trends which are being successfully followed for
the elucidation of the various chemical bases of plant resistance.
The examples include not only those presented in this symposium
book, but others recently reported.

Origins and Actions of Allelochemicals

Identification and Roles of Plant Resistance Compounds including
Phytoalexins. Not all expressions of plant resistance are chemically mediated, but of those that are, the so called secondary
plant compounds appear to be dominant. Among those shown to
mediate resistance are the various classes of phenolic compounds
including flavonoids and aromatic acids, terpenoids including the
sesquiterpene lactones and heliocides, nitrogenous compounds
including amino acids and amides, proteinaceous compounds
including protease inhibitors, glycosidase inhibitors, and phyto-
hemagglutinins, the lectins, toxic seed lipids including unusual
fatty acids, acetylenic and allenic lipids, the fluorolipids and
cyanolipids, the saponins, the tannins, and lignins (1-4). The
roles of these compounds are difficult to elucidate for several
reasons. Not all compounds toxic to one insect or pathogen are
toxic to another. In evolutionary time, some pests have evolved
mechanisms to detoxify, tolerate, or even utilize compounds in
plants from which they derive energy and/or other resources
whereas others do not. Also, several resistance mechanisms may be
operative in the plant. To the extent that biosynthesis of these
compounds is an expression of genetic information, the elucidation
of these compounds and their roles can provide a guide to selections by plant breeders. The description in this book of
several polyhydroxyl alkaloids from legumes that act as glycosidase inhibitors in insects and other animals (Fellows et al., 5)
is an unique example. Byers (6) found that 3-nitropropionic acid
from crown vetch was a feeding deterrent, and also toxic to some
insects and non-ruminant animals. Although this compound serves
as a protectant against insects, its decrease or removal could
improve the use of crown vetch as a forage for some animals.
Cutler et al. (7) reported that several diterpenes and sucrose
esters were inhibitory to etiolated wheat coleoptiles and micro-
bes, but oviposition stimulants for the tobacco budworm, Heliothis

virescens, while combinations of duvanes and sucrose esters inhibited budworm larval development. Thus, multiple, divergent roles
are evident. Harborne (8) and others have shown that phytoalexins
are natural fungitoxic substances produced de novo in plants as a
dynamic response to microbial infection. As many as 25 structures
may be formed in a single interaction. The compounds which
reflect a wide diversity including those listed above have been
isolated from a large number of plant families, generally in
response to infection of the plant by an appropriate microorga-
nism. Bailey (9) reported the biosynthesis of phytoalexins by
plants in response to infection by microorganisms and viruses, but
also in response to a number of chemicals including carbohydrates,
glycoproteins, and peptides isolated from microorganisms. He
reported that the formation of phytoalexins involves the coordinated synthesis of several enzymes, and appears to be mediated
in living cells by metabolites released from adjacent, dying
cells. Wain (10) has found that the phytoalexins often occur in
disease free plants, but build up in response to infection, or
exist in a conjugated form, and are released when the plant is
infected. Also, infection may cause the activation of enzymes,
often oxidases that mediate the production of phytoalexins.
Mechanical and heat treatments have also been shown to promote
phytoalexin synthesis. Gustine (11) reported that the biosynthesis of the major phytoalexin, medicarpin, in ladino clover is stimulated by bacteria, fungi, and certain chemicals. Studies with
sulfhydryl reagents suggest medicarpin accumulation is regulated
by SH-containing components in the cell membrane.

Location and Differentiation of Secondary Constituents Within
Plants. For greatest effect, the allomone should be located at
the appropriate concentration in the plant tissue which is
attacked by the pest at the most vulnerable stage in the pest's
life cycle. The content of an allomone may be similar in ground,
whole plant samples of susceptible and resistant varieties, but
considerably different in specific tissues. The content at the
surface may be different in some situations, but this difference
may not be noted if the whole plant tissue is homogenized and analyzed. For example, cuticular components of tobacco, Nicotiana
tabacum (L.), leaf have been reported by Cutler, et al. (12) to
possess various antibiotic properties. In cotton, Gossypium spp.,
the germ plasm pool is now large enough to select for allelochemicals that are concentrated chiefly in specific tissues. For
example, the promising red marker with demonstrated antibiotic
properties, crysanthemin (13), could be located almost exclusively
in the calyx or the bracts with no loss in impact on pests. Up
until now, the red cottons have not been grown extensively because
they yield about 10% less than isogenic green lines, presumably
because of the biosynthetic costs to the plant of producing the
anthocyanin in all of its surface tissues. Precision in locating
allelochemicals in the preferred feeding sites of insects could
retain resistance properties without concomitant yield decreases
(14).
        Gregory, et al. (15) reported on potato glandular trichome
chemistry as it affects several insects. The trichomes exude a

sticky substance, then sesquiterpenoids are released which cause
enhanced insect movement. Subsequently, the trichomes release
polyphenol oxidase (PPO) and phenol substrates which react to form
quinones. The resultant effect on the insect is cessation of
feeding and death. The glandular trichomes thus contribute to a
concentrated, evidently cost effective, defense system.

Stipanovic (16) reviewed the literature on the function and
chemistry of plant trichomes and glands from a number of plants,
and their relation to insect resistance. There is a great diversity of allelochemicals found in trichomes, glands, nectaries;
chiefly alkaloids, flavonoids, terpenes, and phenols.

Rodriguez (17) also has found trichomes to be important
storage structures for terpenoids, alkaloids, phenols and amines
in the arid land plants that have been the focus of his research.
He and his associates have isolated several chromenes, prenylated
quinones, and sesquiterpenes esterified with phenolic acids which
have repellent and/or cytotoxic activity.

Multiple Factor Contributions. There is increasing evidence for
multiple factor contributions to plant resistance. In laboratory
studies, it has been shown that the growth of tobacco budworm larvae is retarded by a number of compounds isolated from the cotton
plant including gossypol and related compounds, several flavonoids, catechin condensed tannins, cyanidin, delphinidin, and
their glucosides (13, 14, 18, 19). Field studies have shown that
varieties high in several components possess the greatest
resistance (20). Thus, there is the potential to breed for
varieties high in several components, the biosynthesis of which
may be controlled by separate genes. Improved gene insertion
techniques should eventually be applicable to these objectives.
Multiple factor resistance may be less susceptible to adaptation
by pests. Cutler et al. (7) have reported that combinations of
diterpene duvanes and sucrose esters of C_3-C_7 fatty acids are oviposition stimulants for the tobacco budworm, suggesting that a
profile of components is required. Saxena (21) in his studies on
rice plant allelochemicals as oviposition deterrents for several
insects has implicated the importance of volatiles, amino acids,
and trans-aconitic acid, again indicating the requirement for a
profile of components. Kennedy (22) also has shown the requirement for multiple components; phenols, alkaloids, and methyl ketones, in resistance of tomato plants, Lycopersicon, to several
insect pests.

Secondary Metabolites from Higher Plants as Pest Control Agents
An activity related to various efforts to identify compounds in
plants with resistance to pests, and subsequently to attempt selection of lines with higher concentrations, is the utilization of
these compounds (or the synthetic species) in field pest control.
The following includes a few examples from a very large scientific
activity. Jacobson (23) has tested azadirachtin, a sesquiterpenoid
from the neem tree, Azadirachta indica (A. Juss), against several
insects and nematodes. It is effective at concentrations as low as

0.1 ppm, and appears safe for fish, animals and crop plants. The insect feeding deterrents isolated from semiarid and arid land plants by Rodriguez (17) may also find eventual commercial application. Whitehead (24) has reported on the prognosis for using natural products and their analogues for combating pests in Africa. He has suggested that stored grains may be protected from pests with thin coatings of vegatable oils, often smoked, in situations where sophisticated, often expensive procedures are not feasible. Cutler (25) and associates have identified a number of secondary metabolites that have allelochemical effects in plants, fungi, bacteria, and vertebrates. Carboxyatractyloside, isolated from the plant Xanthium strumarium, is both a hypoglycemic and phytotoxic agent. It also has fungistatic activity. The selective action of this compound suggests that it might be developed as a herbicide to control grasses in corn.

Photosensitizers, Elicitors, and Primitive Immune Systems. Plants synthesize chemicals which possess a wide range of biological activities. Natural products that require light for expression of their toxic biological consequences are among the most interesting of these biological activities. Such photosensitizers or photo-toxins are structurally variable, are produced by widely divergent flowering plant families, and mediate broad-spectrum toxic reactions. Alkaloids, acetophenones, furocoumarins, furochromes, and polyines are among the classes of compounds identified by Downum (26) and associates.

Arnason et al. (27) have reported on other photosensitizers; polyacetylenes, β-carbolines, and extended quinones with high activity to insects. α-Terthienyl was found to be more toxic to mosquito than DDT. The toxicity of some of these compounds is mediated by the production of singlet oxygen although free radicals may be involved. Furanocoumarins and furanoquinolines, on the other hand, interact with DNA. The furanocoumarin, 8-methoxy-psoralen, was found highly toxic to Spodoptera larvae. Kogan and Paxton (28) reported that there are a number of natural inducers of plant resistance to insects such as climatic factors, pesticides, growth regulators, and microbial infections. They concluded that resistance of plants to arthropods generally resulted from genetically controlled, injury-independent accumulation of allomonal agents, whereas for many pathogens, phytoalexin formation is the result of a post-challenge response. However, they provided several examples where herbivore attacks did elicit resistance in plants, an example of which is the role of phytoalexins as antiherbivory agents in soybeans against the Mexican bean beetle, Epilachna varivestis.

An example of a so-called primitive immune system, somewhat parallel to that in animals, is provided by the finding by Nelson et al. (29) that two proteinase inhibitors accumulate in leaves of tomato plants when attacked by chewing insects or mechanically wounded. The accumulation of the two antinutrient proteins is apparently a defense response and is initiated by the release of a putative wound hormone called the proteinase inhibitor inducing

factor. A second wound at nine hours later triples the steady
rate of inhibitor accumulation.

Phytochemical Disruption of Hormonal Processes. Hormones are unique regulators of insect and plant growth and other life processes.
Allelochemicals which operate by disrupting hormonal processes
can be visualized as important components of plant defensive
mechanisms. Bowers (30) has reviewed the elucidation by himself
and others of compounds identified as disrupting hormonal processes in insects. Juvabione, isolated from balsam fir, induced
the linden bug and other insects to molt into nymphal-adult intermediates or supernumerary nymphs which eventually died without
completing metamorphosis. Several other compounds isolated from
protected plants or trees with quasi-juvenile hormone activity
were also identified including farnesol, dehydrojuvabione, sesa-
min, sesamolin, and the juvocimones. Molting and differentiation
in insects are controlled by molting hormones called ecdysones.
Subsequently, a structurally similar polyhydroxysteroid,
ponasterone A, was isolated from the plant Podocarpus nakaii and
demonstrated to have molting hormone activity. This has led to
the identification of nearly 70 additional plant ecdysteroids with
molting hormone activity. Ecdysteroids have been shown to have
anti-insect activities, and were thus assumed to play a role in
plant defensive strategy. Bowers (30) and coworkers have also
discovered that certain plants contained compounds which initiated
behavioral responses in the American cockroach, identical to those
induced by the natural sex pheromone of the virgin female. The
active principle in several evergreen species was isolated and
found to be D-bornyl acetate. Later, Germacrene D was isolated
from the plant Erigeron annus and shown also to have sexual activity for this insect. Subsequently, the true pheromone of this
insect was isolated from the midgut and shown to be Periplanone B,
a derivative of Germacrene D. It was speculated that the plant's
biosynthesis of these compounds may disrupt pheromonal communication of insects that feed on the plant.

Biorational Approaches to the Development of Pest Control Agents.
As the roles of natural products in the control of pests has
increased, there has been a concomitant growth in the understanding of biochemical processes. Knowledge of metabolism,
biosynthetic processes, neurochemistry, regulatory mechanisms, and
many other aspects of plant, animal, and pest biochemistry has
provided a more complete basis for understanding the modes of
action of pest control agents including allelochemicals. The
exploitation of knowledge of the chemical bases of biological
interactions (i.e., biorational approaches) may lead to the
recognition and utilization (i.e., synthesis, chemically directed
selections, gene insertions) of new molecules designed to act at a
particular site or to block a key step in a biochemical process.
     Attempts to develop resistant lines or cultivars by
employing analyses to determine whether correlations between the
allelochemical and resistance are statistically significant, has
only recently been explored on a substantive scale. Efforts to
make selections based on the high (or low) concentration of an
allelochemical depend on the availability of a large, diverse germ

plasm pool that contains entries from many countries, both from
temperate zones and particularly from the tropics where they have
endured severe challenges by pests. Subsequently, chemical and
biochemical evaluations may provide guidance as breeders make
crosses to develop resistant lines with required commercial attri-
butes. This approach is now being initiated at several laboratories including ours at Mississippi State where we have monitored
20 cotton Gossypium spp. lines (13, 14) and 15 corn Zea mays L.
lines (31) for allelochemicals. The results have formed the basis
for subsequent selections. As biotechnological techniques improve
so that gene traits for allelochemicals can be transferred to crop
plants, the biorational approach will aid, and perhaps control,
the gene selection process.

Genetic Manipulation of Plants for Resistance. Biotechnology
promises to play a significant role in crop improvement and pro-
ductivity in the coming years. The improvement of crop plants
using recombinant DNA techniques will require the development of
transformation vectors and regeneration technology. With the
development of these techniques, identification of the genes to be
transferred will be required. Marvel (32) has recently reviewed
the prognosis for biotechnology in crop improvement. Kennedy (22)
has observed that although there are a number of biologically
active chemicals which have potential for genetic manipulation,
the manipulations are likely to be complex in that the actual
level of resistance often depends upon the larger chemical context
within which that compound occurs in the plant. A further compli-
cating factor is that the introduction of a particular
biochemically-mediated resistance to one insect species may have
unanticipated and undesirable effects on other non-target insects.
A thorough understanding of the biochemical interactions will be
required to develop plants with high levels of allelochemical by
mediated resistance. The potential for, (and some disadvantages
of) genetic manipulation is illustrated by investigations of
Rosenthal (33) and Dahlman and Berge (34) with L-Canavanine, the
non-protein amino acid biosynthesized by Dioclea megacarpa and
other legumes. L-canavanine, in this instance, exists in the free
form, but is incorporated by insects into their protein whence
toxic effects occur. In another scenario, L-canavanine might be
inserted genetically into plant protein, and thus exert its toxicity. However, if canavanine containing proteins were also toxic
to mammals, the purpose would be defeated. In any event, there
are large numbers of allelochemicals from which selections for
insertion might be made. Of course, the ecological and agronomic
fitness of the transformed plant will have to be evaluated, and it
may take several years to produce a commercially promising cultivar.

Mechanisms of Plant-Pest Interactions

Coevolution of Plants and Pests. Plants and their herbivores and
pathogens (including pests) coevolve, and the continuing adjust-
ments of one to the other reflect the biosynthesis of defensive
compounds by the plant and the development of detoxification

mechanisms by the pest. The dynamic nature of this relationship is illustrated by the ability of insects to induce detoxifying enzymes within 24 h. when challenged by a toxic agent. Plant injury, in turn, can elicit the biosynthesis of additional quantities of resistance agents. Cates (35) in a study of Douglas-fir/budworm interactions showed changing responses of one to the other as a consequence of plant toxins, tree nutritional levels, geographical location, and climate stresses. Fellows et al. (5) and others identified several polyhydroxy alkaloids in higher plants which structurally resemble monosaccharides and are potent inhibitors of glycosidase activity in insects, animals, and microorganisms. Because these compounds also inhibit trehalose, and trehalose is an important storage carbohydrate in insects, incorporation of the alkaloids into the legume pods on which bruchid beetles (Callosobruchus spp.) feed could provide a defense mechanism. However, the toxicity to other strains of the bruchid beetle, and to other insect species differ, suggesting that adaptations have occurred.

Biochemical Mechanisms. The role of natural products in the control of pests has increased in recent decades as the chemist has acquired more sophisticated tools with which to elucidate complex structures. This, in turn, has led to explosive growth in the understanding of biochemical processes. Knowledge of metabolism, biosynthetic processes, neurochemistry, regulatory mechanisms, and many other aspects of plant, animal, and insect biochemistry has provided a more complete basis for understanding the mechanisms of plant resistance.

Bell et al. (36) reported on the biochemistry of cotton resistance to bacterial and fungal infections. Cotton plants synthesize both sequiterpenoid phytoalexins and proanthocyanidins in response to microbial infections. From their continuing research, they have identified a large number of terpenoid compounds, associated with gossypol, and have suggested the mechanisms by which they are biosynthetically formed from one another. Gustine (11), using sulfhydryl reagents, showed that medicarpin accumulation is regulated by SH-containing components in the cell membrane. Fellows et al. (5) elucidated the mechanism by which glycosidases from insects, mammals, and microorganisms are inhibited by polyhydroxyalkaloids, and their possible ecological role. Gregory et al. (15) showed that an oxidation reaction, initiated upon disruption of potato trichomes, between a unique polyphenoloxidase enzyme and phenolic substrate(s), resulted in quinone formation and subsequent insect immobilization.

Nutritional Antagonisms. Some plants possess the capability to present a nutritionally undesirable diet to pests apart from their biosynthesis (or failure to biosynthesize) of allelochemicals. Rosenthal (33) and Dahlman and Berge (34) have shown that L-Canavanine, a non-protein amino acid of certain leguminous plants, is incorporated into proteins of some insects with toxic results. The tobacco hornworm is sensitive to the formed protein, however the tobacco budworm, Heliothis virescens (Fab.), and the bruchid beetle, Caryedes brasiliensis, are able to tolerate and

metabolize high concentrations. The biochemical detoxification
routes have been elucidated.

Reese (37) has studied nutrient-allelochemical interactions
using diets supplemented by plant extracts. He has concluded that
many of the deleterious effects may be due primarily to various
interactions between these allelochemical and essential nutrients.
It is important to not only consider the presence of nutrients,
but also the bioavailability of these nutrients to the phytophagous insect. He cited findings where neonate larvae were found
more sensitive to plant allelochemicals than older larvae, and
interpreted the greater resistance of the older larvae in terms of
increasing capability to utilize available nutrients, probably
because of the induction of enzymatic detoxification systems.
Blum (38) observed that many insects have the capability to feed
on "so-called" toxic plants, having developed the required detoxification systems. Nutritional factors may become critical as a
consequence of stresses to the insect involved with the energy
requirements devoted to detoxification, deactivation, or
sequestration.

Finally, there are general nutritional requirements for
essential amino acids, fatty acids, sterols, and vitamins by
pests. The absence or deficiency of one or more of these may
limit growth or development.

Differential Sensory Perceptions of Plant Compounds by Insects.
These perceptions are a function of the chemical composition of
the plant and the chemical sensory (neurological) resources
(including the brain) of the insect. The result of these interactions is a behavioral response. The understanding that not all
insects respond identically to a stimulus brings diversity to
these investigations. Narahashi (39) and others have established
that insects may possess several neurochemical targets. In pheromone perception by insects, at least two different receptor cells
have been found to be involved. These receptors are sufficiently
flexible to perceive more than one compound and are sensitive to
the precise ratios of compounds. Receptor specificity is further
demonstrated in the differential responses to chiral forms of
pheromones and in the responses from receptors on the antennae of
males and females. Visser (40) has studied insect perception of
plant characters with regard to selection of a plant for feeding.
By means of gustatory receptors, an insect is informed about the
plant's nutritional quality. Plant odors are the chemical
messengers for the insect's orientation. The differential perception of green odor, being composed of C-6 alcohols, aldehydes, and
derivative acetates is a common feature in the sensitivity and
selectivity to phytophagous insects with the sensitivity of olfactory receptors for the individual components in the complex
varying. Other compounds also are involved in host plant selection.

Psychomanipulation of Insects. Murdock et al. (41) have developed
the hypothesis that certain plant chemicals consumed by feeding
insects interfere with information processing in the insect's
central nervous system and thereby modify behavior. Two plant

components, nicotine and lobeline, act as antagonists of the nico-
tinic cholinergic receptor. Gamma-aminobutyric acid (GABA) receptors are blocked by two chemicals of plant origin, bicucculine and
picrotoxin. Scopolamine is an excellent blocking agent for
muscarinic cholinergic receptors. Physostigmine, another active
plant chemical that is a synaptic clearance antagonist, is also a
strong cholinesterase inhibitor. This inhibition results in
intensified stimulation of the postsynaptic cell. A number of
similar effects have been established with plant chemicals. The
overall effect is that plants obtain a measure of protection from
insects by accumulating secondary substances which act in the CNS
to change the insect's behavior in ways that reduce or prevent
further feeding such as suppressing appetite, interfering with CNS
integrative procedures, blocking learning or memory, or distorting
vision.

Summary. This review has been an attempt to identify developing
trends by which research to elucidate the chemistry of plant
resistance to pests is evolving. Although the impact of chemical
knowledge on the development of resistant varieties has been
hitherto limited, developing techniques and insights concerning
these and other approaches may now make a much larger, more
vigorous contribution, even a direction, to this activity.

Acknowledgments and Disclaimers

Mention of a trademark, proprietary product or vendor does not
constitute a guarantee or warranty of the product by the U.S.
Department of Agriculture and does not imply its approval to the
exclusion of other products or vendors that may also be suitable.
       This chapter was prepared by a U.S. Government employee as
part of his official duties and legally cannot be copyrighted.

Literature Cited

1.   Wallace, J. W.; Mansell, R. L. "Biochemical Interaction
      Between Plants and Insects"; Recent Advances in
      Phytochemistry, 1976; Vol. 10, Plenum Press, N.Y., 425 pp.
2.   Rosenthal, G. A.; Janzen, D. H. "Herbivores: Their
      Interaction With Secondary Plant Metabolites"; Academic
      Press, N.Y.; 1979, 718 pp.
3.   Hedin, P. A. "Host Plant Resistance to Pests"; ACS SYMPOSIUM
      SERIES No. 62, American Chemical Society, Washington, D.C.,
      1977, 286 pp.
4.   Hedin, P. A. "Plant Resistance to Insects"; ACS SYMPOSIUM
      SERIES No. 208, American Chemical Society, Washington, D.C.,
      1982, 375 pp.
5.   Fellows, L. E.; Evans, S. V.; Nash, R. J.; Bell, E. A.   Ch.
      6, In Natural Resistance of Plants: Roles of Allelochemicals;
      Green, M. B; Hedin, P. A., Eds.; ACS SYMPOSIUM SERIES, American
      Chemical Society, Washington, D.C., 1986.
6.   Byers, R. A.; Gustine, D. L.; Moyer, B. G.; Bierlein, D. L.
      Ch. 8, In Natural Resistance of Plants: Roles of
      Allelochemicals ; Green, M. B.; Hedin, P. A., Eds.; ACS

SYMPOSIUM SERIES, American Chemical Society, Washington,
D.C., 1986.

7. Cutler, H. G.; Severson, R. F.; Cole, P. D.; Jackson, D.M.
    Ch. 15, In Natural Resistance of Plants: Roles of
    Allelochemicals; Green, M. B.; Hedin, P. A., Eds.; ACS
    SYMPOSIUM SERIES, American Chemical Society, Washington,
    D.C., 1986.

8. Harborne, J. B. Ch. 3, In Natural Resistance of Plants to
    Pests: Roles of Allelochemicals; Green, M. B.; Hedin, P. A.,
    Eds.; ACS SYMPOSIUM SERIES, American Chemical Society,
    Washington, D.C., 1986.

9. Bailey, J. A. "Abstracts of Papers"; 189th National Meeting
    of the American Chemical Society, Miami Beach, Fl., April
    1985; American Chemical Society, Washington, D.C.; PEST 10.

10. Wain, R. L. Ch. 2, In Natural Resistance of Plants to Pests:
    Roles of Allelochemicals; Green, M. B.; Hedin, P. A., Eds.;
    ACS SYMPOSIUM SERIES, American Chemical Society, Washington,
    D.C., 1986.

11. Gustine, D. L. Ch. 5, In Natural Resistance of Plants to
    Pests: Green, M. B.; Hedin, P. A., Eds.; ACS SYMPOSIUM
    SERIES, American Chemical Society, Washington, D.C., 1986.

12. Cutler, H. G.; Reid, W. W.; Deletang, J. Plant Cell Physiol.
    1977, 18, 711.

13. Hedin, P. A.; Jenkins, J. N.; Collum, D. H.; White, W. H.;
    Parrott, W. L.; MacGown, M. W. Experientia 1983, 39, 799.

14. Hedin, P. A.; Jenkins, J. N.; Collum, D. H.; White, W. H.;
    Parrott, W. L. Ch. 20, In "Plant Resistance to Insects";
    Hedin, P. A., Ed.; ACS SYMPOSIUM SERIES No. 208, American
    Chemical Society, Washington, D.C. 1982. p. 347.

15. Gregory, P.; Tingey, W. M.; Ave, D. A., Bouthyette, P. Y.
    Ch. 13, In Natural Resistance of Plants to Pests: Roles of
    Allelochemicals; Green, M. B.; Hedin, P. A., Eds.; ACS
    SYMPOSIUM SERIES, American Chemical Society, Washington,
    D.C., 1986.

16. Stipanovic, R. D. Ch. 5, In "Plant Resistance to Insects";
    Hedin, P. A., Ed.; ACS SYMPOSIUM SERIES No. 208, American
    Chemical Society, Washington D.C., 1982, p. 69.

17. Rodriguez, E. Ch. 31, In "Bioregulators for Pest Control";
    Hedin, P. A., Ed.; ACS SYMPOSIUM SERIES No. 276, American
    Chemical Society, Washington, D.C. 1985, p. 447.

18. Bell, A. A.; Stipanovic, R. D. Proc. Beltwide Cotton
    Production Res. Conf. 1977, Jan. 10-12, Atlanta, Ga., 244.

19. Chan, B. G.; Waiss, A. C.; Binder, R. G.; and Elliger, C. A.
    Entomologia Exp. Appl. 1978, 24, 94.

20. Jenkins, J. N., unpublished data.

21. Saxena, R. C. Ch. 12, In Natural Resistance of Plants to
    Pests: Roles of Allelochemicals; Green, M. B.; Hedin, P. A.,
    Eds.; ACS SYMPOSIUM SERIES, American Chemical Society,
    Washington, D.C., 1986.

22. Kennedy, G. G. Ch. 11, In Natural Resistance of Plants to
    Pests: Roles of Allelochemicals; Green, M. B.; Hedin, P. A.,
    Eds.; ACS SYMPOSIUM SERIES, American Chemical Society,
    Washington, D.C., 1986.

23. Jacobson, M. Ch. 18, In Natural Resistance of Plants to Pests; Green, M. B.; Hedin, P. A., Eds.; ACS SYMPOSIUM SERIES, American Chemical Society, Washington, D.C., 1986.

24. Whitehead, D. L. Ch. 29, In "Bioregulators for Pest Control"; Hedin, P. A., Ed.; ACS SYMPOSIUM SERIES No. 276, American Chemical Society, Washington, D.C. 1985, p. 409.

25. Cutler, H. G. Ch. 32, In "Bioregulators for Pest Control"; Hedin, P. A., Ed.; ACS SYMPOSIUM SERIES No. 276, American Chemical Society, Washington, D.C. 1985, p. 455.

26. Downum, K. R. Ch. 16, In Natural Resistance of Plants to Pests; Green, M. B.; Hedin, P. A., Eds.; ACS SYMPOSIUM SERIES, American Chemical Society, Washington, D.C., 1986.

27. Arnason, T.; Towers, G. H. N.; Philogene, B. J. R.; Lambert, J. D. H. Ch. 9, In "Plant Resistance to Insects"; Hedin, P. A., Ed.; ACS SYMPOSIUM SERIES No. 208, American Chemical Society, Washington, D.C., 1982, p. 139.

28. Kogan, M.; Paxton, J. Ch. 9, In "Plant Resistance to Insects"; Hedin, P. A., Ed.; ACS SYMPOSIUM SERIES No. 208, American Chemical Society, Washington, D.C., 1982, p. 153.

29. Nelson, C. E.; Walker-Simmons, M.; Makus, D.; Zuroske, G.; Graham, J.; Ryan, C. A. Ch. 6, In "Plant Resistance to Insects"; Hedin, P. A., Ed.; ACS SYMPOSIUM SERIES No. 208, American Chemical Society, Washington, D.C., 1982, p. 103.

30. Bowers, W. S. Ch. 15, In "Bioregulators for Pest Control"; Hedin, P. A., Ed.; ACS SYMPOSIUM SERIES No. 276, American Chemical Society Washington, D.C., 1985, p. 225.

31. Hedin, P. A.; Davis, F. M.; Williams, W. P.; Salin, M. L. J. Agr. Food Chem. 1984, 32, 262.

32. Marvel, J. T. Ch. 34, In "Bioregulators for Pest Control"; Hedin, P. A., Ed.; ACS SYMPOSIUM SERIES No. 276, American Chemical Society, Washington, D.C., 1985, p. 477.

33. Rosenthal, G. A. Ch. 16, In "Plant Resistance of Insects"; Hedin, P. A., Ed.; ACS SYMPOSIUM SERIES No. 208, American Chemical Society, Washington, D.C., 1982, p. 279.

34. Dahlman, D. L.; Berge, M. A. Ch. 10, In Natural Resistance of Plant to Pests: Roles of Allelochemicals; Green, M. B.; Hedin, P. A., Eds.; ACS SYMPOSIUM SERIES, American Chemical Society, Washington, D.C., 1986.

35. Cates, R. G. Ch. 9, In Natural Resistance of Plants to Pests: Roles of Allelochemicals; Green, M. B.; Hedin, P. A., Eds.; ACS SYMPOSIUM SERIES, American Chemical Society, Washington, D.C., 1986.

36. Bell, A. A.; Mace, M. E.; Stipanovic, R. D. Ch. 4, In Natural Resistance of Plants to Pests: Roles of Allelochemicals; Green, M. B.; Hedin, P. A., Eds.; ACS SYMPOSIUM SERIES, American Chemical Society, Washington, D.C., 1986.

37. Reese, J. C. Ch. 13, In "Plant Resistance to Insects"; Hedin, P. A., Ed.; ACS SYMPOSIUM SERIES No. 208, American Chemical Society, Washington, D.C., 1982, p. 231.

38. Blum, M. S. Ch. 15, In "Bioregulators for Pest Control"; Hedin, P. A., Ed.; ACS SYMPOSIUM SERIES No. 276, American Chemical Society, Washington, D.C., 1985, p. 265.

39. Narahashi, T. "Neurotoxicology of Insecticides and Pheromones", Narashi, T., Ed.; Plenum Press, N.Y., 1979.
40. Visser, J. H. Ch. 12, In "Plant Resistance to Insects"; Hedin, P. A., Ed.; ACS SYMPOSIUM SERIES No. 208, American Chemical Society, Washington, D.C., 1982, p. 215.
41. Murdock, L. L.; Brookhart, G.; Edgecomb, R. S.; Long, T. F.; Sudlow, L. Ch. 24, In "Bioregulators for Pest Control"; Hedin, P. A., Ed.; ACS SYMPOSIUM SERIES No. 276, American Chemical Society, Washington, D.C., 1985, p. 337.

RECEIVED October 16, 1985

ROLES OF PHYTOALEXINS IN PLANT–PLANT INTERACTIONS

2

Some Chemical Aspects of Plant Disease Resistance

R. L. Wain

Department of Chemistry, University of Kent, Canterbury, England

Growing plants are exposed to pressures of environment and disease, and without built-in defense mechanisms, they could not survive. They protect themselves against stresses such as drought by producing a chemical which reduces both the rate of growth and leaf transpiration. Plants can also exert chemical defenses against fungal pathogens. Some of the chemicals which operate occur normally in plants; others build up in the tissues following an attempted invasion by the fungus. Like chlorophyll, all these chemicals have arisen over thousands of years of plant development, and unlike manmade systemic fungicides, they never lose their effectiveness. Their presence in plants helps to explain the wide-spread natural resistance shown towards disease. Isolation and identification of these naturally occurring fungicides is now a routine procedure and such chemicals from a resistant plant can sometimes be used to protect a susceptible species from disease attack.

Whereas animals can take shelter, plants remain in the same place, no matter how unfavourable environmental conditions may be. It is not surprising then that throughout their evolutionary history, plants have developed their chemistry to protect themselves against environmental stresses such as drought, waterlogged soil, and soil salinity. There is now evidence that they do this by biosynthesising relatively large amounts of the endogenous plant-growth hormone inhibitor abscisic acid. This then restricts the activities of

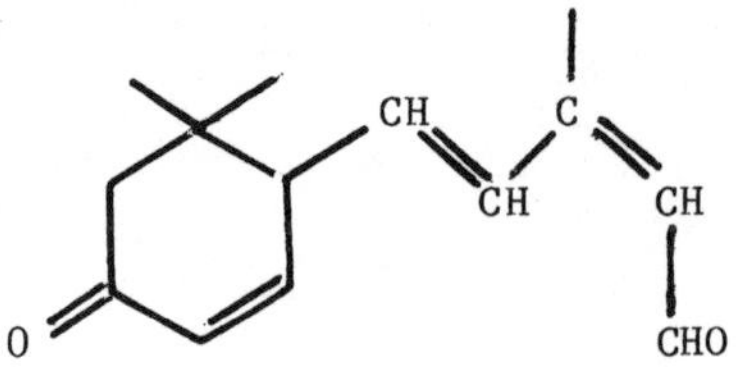

Abscisic acid

0097-6156/86/0296-0016$06.00/0

the growth hormones which are responsible for promoting extension
growth and cell division in plants. In this way, growth may virtually cease, thereby conserving the plant's energy during the critical stress period. The marked increase in the abscisic acid level
which can occur during water stress was first observed by Wright in
my laboratory (1,2). In a typical experiment, the abscisic acid
level in the leaves of a tomato plant was shown to increase over
fifty fold when water was withheld for three days. It was later
shown elsewhere that not only does this abscisic acid restrict the
plant's growth but its presence in the leaves leads to a closure of
the stoma so that water loss by transpiration is decreased (3).

Turning now to plant diseases, we know that although all plants
are exposed to a very wide range of potentially pathogenic fungi,
they are completely resistant to most of them. Indeed it is true
that in general, under natural conditions, resistance is the rule.
There are many ways in which a plant can resist fungal infection,
but it is now well established that amongst these, chemical defense
mechanisms are important.

That fungicidal compounds are present in the washings of
healthy leaves of a number of plant species was demonstrated by
Topps and myself (4). Sakurai, also working in my laboratory, has
shown that an acidic fungicidal substance, not yet identified, can
be obtained from the washings of healthy leaves of Vicia faba. Such
compounds then could be a factor in determining whether the spores
of some fungi can germinate on the leaf surface.

A well recognised defense mechanism against a potential pathogen is the plant cuticular layer which may provide a barrier to
infection to certain fungi. If, however, a germinating fungal
spore does penetrate into the plant cells, it must find suitable
nutritional and other conditions for successful establishment. In
some cases, this establishment may fail owing to the rapid death of
the host tissue cells at the site of infection so that here, hyper-
sensitivity of the host plant could be a primary cause of
resistance.

Research in recent years has shown that protective chemicals in
plants can operate in disease resistance. Some of these chemicals
occur in disease-free plants, but fungicidal compounds can build up
in tissues in response to attempted infection. When this occurs,
these defense chemicals are known as phytoalexins. Papers to be
presented at this conference by Cutler et al., Harborne, Bailey,
Bell et al. and Gustine will deal with recent developments in
research on phytoalexins.

Earlier research on phenolic substances in relation to plant
disease resistance was concerned not only with the possible protective effects of preformed phenolic compounds, but also whether these
compounds are mobilized or their synthesis is promoted at the site
of infection. Another aspect which has received attention is that
the phenolic compound might be liberated from its glycoside or sugar
ester at the infection site in response to infection (5). Chemical
changes in infected tissue due to the activity of polyphenoloxidases and peroxidases, leading to the production of quinones and
other fungicidal compounds, have also been investigated (6).

The possibility that fungicidal action may arise from the
liberation of an aglycone from its glycoside within the tissues of
the fungus provides an approach to systemic fungicides which is

being followed in my laboratory. Well known synthetic fungicides, not themselves systemic, are converted to their more water soluble β-D-glucosides which are more able to move within the plant. Liberation and metabolism of the sugar from the glucoside by the plant is thought to be unlikely since adequate carbohydrate is available from the phytosynthetic pathway. The pathogen, however, which is entirely dependent on supplied carbohydrate, might well adapt itself to hydrolyze the glycoside, thereby releasing the fungicidal aglycone.

The literature on phytoalexins is now very extensive as Dr. Harborne will relate in his paper. These fungitoxic substances which are produced in plants in response to attempted fungal invasion are thought to arise from an interaction between specific metabolic systems of the host and the fungus. But although phytoalexins are synthesised and accumulated by plant tissues in response to the presence of fungi, phytoalexin formation can be stimulated by other means. Thus, chemical, mechanical and heat treatment, and anaerobic storage of host plant tissues (7,8) have all been shown to promote the synthesis of phytoalexins. All these "stress" factors then, both biological and non-biological, can initiate the changes in host plant metabolism upon which phytoalexin formation depends. This suggests that the production of phytoalexin might arise from a shift in an already existing biosynthetic mechanism; indeed, the initiation of a completely new synthetic pathway involving specific enzymes is most unlikely as it would require the operation of new genes. It would be logical, therefore, to expect traces of the phytoalexin to occur in the normal, uninfected plant together with substances which can yield the phytoalexin in response to the activation or suppression of enzyme reactions which are already established within the host.

Albersheim has developed a concept of phytoalexin production being promoted by specific "elicitors" - that is, by molecules produced by certain fungi, which can sometimes be found in fungal culture media (9). Such elicitors have been isolated and their role in the defensive process has been investigated (10).

The resistance of plants towards disease may well be related to phytoalexin formation, but whatever be the precise function of these compounds, they can only represent one of the complex of factors which operate in disease resistance and immunity. Any protective substances present in the healthy plant must also be important in this connection. A good example of such a substance is provided by our discovery that healthy seedlings of broad bean (Vicia faba) contain a potent antifungal chemical to which we have given the name Wyerone.

Our first indication that Wyerone occurs in broad bean plants arose from an observation that fungal growth on nutrient sugar was

$$\text{C}_2\text{H}_5\text{CH} = \text{CH.C} \equiv \text{C.CO} \underset{\text{O}}{\overset{\underset{c}{}}{\longleftarrow}} \overset{t}{\longrightarrow} \text{CH} = \text{CHCOOCH}_3$$

Wyerone

inhibited by the presence of segments of the stem or root tissue (11). The fungicidal compound present in healthy seedlings was isolated, its structure and steric configuration were established and the substance (Wyerone) was also synthesised (12,13).

Wyerone was found to show a wide fungicidal spectrum when tested against phytopathogens in spore germination tests (14). It showed high in vitro activity, particularly against Alternaria brassicicola (ED$_{50}$ 3 ppm). However, it was some 40 times less active against Botrytis cinerea (ED$_{50}$ 125 ppm). This is of considerable interest since A. brassicicola does not affect broad bean plants whereas Botrytis spp. are the pathogens which cause the well known chocolate spot disease.

Wyerone levels in Vicia faba have been shown to increase in response to attempted fungal invasion so in this sense, Wyerone, although it occurs in healthy tissues must also be considered as a phytoalexin. Wyerone acid has been isolated from bean infected tissues (15) and this fungicidal acid (in which the COOCH$_3$ group of Wyerone has been hydrolysed to −COOH) must also be considered as a substance which protects against disease.

The detection and isolation of phytoalexins and other naturally occurring fungicides has been greatly simplified using the technique whereby plant tissue extracts are subjected to thin layer chromatography after which the developed plate is sprayed with spores of an appropriate fungus suspended in a nutrient solution. On incubating the plates, the darkly pigmented fungus grows over the plate except in the areas where the antifungal compounds are situated. Using this technique, we found another source of naturally occurring fungicides. This came from a speculation on why plant roots growing in soil usually remain healthy although they are exposed to a wide range of bacteria and fungi which can readily destroy dead root tissue. Such considerations led us to investigate the chemicals exuded by pea and bean seedling roots into the water in which the roots were growing. It was found in both cases that antifungal compounds were present (16). Thus it is clear that the living root operates to defend itself in the hostile environment of the soil.

Amongst the known phytoalexins, one finds considerable variations in chemical structure and some of these naturally occurring fungicides taken from one plant have been shown to protect other plants against fungal attack. This has been shown for example with the solanaceous phytoalexin capsidiol which is highly active against Phytophthora infestans. When the compound was applied to tomato plants at 100 ppm prior to inoculation with P. infestans, it provided almost complete protection (17).

Another example of the use of a naturally occurring fungicide in plant protection is provided by 'sclareol' − an epimeric mixture of sclareol and 13-episclareol. We have shown that this mixture of diterpenes which occurs on the surface of healthy leaves of tobacco (Nicotiniana glutinosa) will prevent the germination of rust spores at low concentrations (18). This finding led us to examine its use as a rust fungicide. It was then found that when it was applied as a spray to broad bean and dwarf bean at 100 ppm, it provided almost complete protection against certain rust diseases (19).

These examples illustrate possible agricultural uses for these organic fungicides which occur naturally within the plant kingdom. Such compounds can often be synthesised as has been done, for

example, with pisatin (20) Wyerone (13) rishitin (21) vignafuran
(22) and orchinol (23). Furthermore, the approach can also be used
to prepare analogues closely related to the naturally occurring
fungicide. Not only might these be of agricultural significance,
but they might be of value in controlling fungal pathogens of man
and animals. Fungal infections of the eye, for example, although
not common are extremely difficult to treat with existing fungicides
because these chemicals are often irritant or toxic to the delicates
of the eye. Such cases of keratomycosis might however be more successfully treated with naturally occurring fungicides and these
possibilities are now being investigated.

Literature Cited

1.   Wright, S. T. C. Planta, 1969, 86, 10.
2.   Wright, S. T. C.; Hiron, R. W. P. Nature, Lond., 1969, 224,
      719.
3.   Jones, R. J.; Mansfield, T. A. J. Exp. Bot., 1970, 21, 714.
4.   Topps, J. A.; Wain, R. L. Nature, London, 1957, 179, 652.
5.   Flood, A. E.; Kirkham, D. S. In Phenolics in Plants in Health
      and Disease; J. B. Pridham ed.; Pergamon Press, Oxford, 1960.
6.   Goodman, R. N.; Kiraly, Z.; Zaitlin, M. Biochemistry and
      Physiology of Infectious Plant Disease; van Nostrand Co.,
      Princeton N.J., 1967, p. 187.
7.   Cruickshank, I. A. M.; Perrin, D. R. Aus. J. Biol. Sci., 1963,
      16, 111.
8.   Perrin, D. R.; Cruickshank, I. A. M. Aus. J. Biol. Sci., 1965,
      18, 803.
9.   Ayers, A. R.; Ebel, J.; Valent, B.; Albersheim, P. Plant
      Physiol., 1976, 37, 751; 760; 766.
10.   Ebel, J.; Ayers, A. R.; Albersheim, P. Plant Physiol., 1976,
      37, 775.
11.   Spencer, D. M.; Topps, J. H.; Wain, R. L. Nature, Lond., 1957,
      179, 651.
12.   Fawcett, C. H.; Spencer, D. M.; Wain, R. L.; Jones, Sir Ewart;
      Le Quan, M.; Page, C. B.; Thaller, V. Chem. Comm., 1965, 1,
      422.
13.   Fawcett, C. H.; Spencer, D. M.; Wain, R. L.; Fallis, A. G.;
      Jones, Sir Ewart; Le Quan, M.; Page, C. B.; Teller, V.;
      Shubrook, D. C.; Witham, P. M. J. Chem. Soc., 1968, p. 2455.
14.   Fawcett, C. H.; Spencer, D. M.; Wain, R. L. Neth. J. Plant
      Path., 1969, 75, 72.
15.   Deverall, B. J. Proc. Symp. Phytochem. Soc., 1971, p. 217.
16.   Burden, R. S.; Rogers, P. M.; Wain, R. L. Ann. Appl. Biol.,
      1974, 78, 59.
17.   Ward, E. W. B.; Unwin, C. H.; Stoessel, A. Phytopath., 1975,
      65, 168.
18.   Bailey, J. A.; Vincent, G. G.; Burden, R. S. J. Gen.
      Microbiol., 1974, 85, 57.
19.   Bailey, J. A.; Carter, G. A.; Burden, R. S.; Wain, R. L.
      Nature, Lond., 1975, 255, 328.
20.   Bevan, C. W. L.; Birch, A. J.; Moore, B.; Mukerjee, S. K.
      J. Chem. Soc., 1964, p. 5991.
21.   Katsui, N.; Matsunaga, A.; Imazumi, K.; Matsume, T.; Tomiyama,
      K. Tetrahedron Lett., 1971, 983.

22.   Preston, N. W.; Chamberlain, K.; Skipp, R. A. <u>Phytochem.</u>,
       1975, 14, 1983.
23.   Stoessel, A.; Rock, G. L.; Fisch, M. H. <u>Chem. Ind.</u>, 1974,
       p. 703.

RECEIVED October 9, 1985

3

The Role of Phytoalexins in Natural Plant Resistance

Jeffrey B. Harborne

Plant Science Laboratories, University of Reading, Reading, RG6 2AS, United Kingdom

Phytoalexins are natural fungitoxic substances produced de novo in plants as a dynamic response to
microbial infection. An astonishing range of structures may be formed, as many as 25 substances in a
single interaction. They are formed consistently
within a given plant group; in a survey of 60 species
of the legume tribe Vicieae, every member tested
responded. Although phytoalexins have been most widely
studied in the Leguminosae, they have been recorded in
over 20 other families. These plants range from
trees, through shrubs to herbs and from monocots to
dicots. There may be limits to the distribution of
this defense mechanism and results of exploring this
response in further families will be presented. The
range of chemical structures encountered will be
reviewed and their role in protecting plants from
microbial infection will be assessed.

The concept that plants are capable of responding rapidly to herbivory by becoming increasingly unpalatable is a relatively new one.
Only quite recently has experimental evidence been obtained indicating chemical changes in the leaves of certain trees as a consequence of insect feeding (1). By contrast, experiments indicating
that plants are capable of synthesizing fungitoxic substances de
novo to ward off microbial infection were carried out around the
turn of the last century (2). The modern development of the phytoalexin concept of Müller and Borger (3) can, however, be precisely
dated to 1960. In that year, Cruickshank and Perrin (4) identified
the first phytoalexin, pisatin, a pterocarpan from infected pea
pods. Since then, phytoalexin research has been one of the dominant
themes of physiological plant pathology. Chemical collaboration has
been an essential component since many of the phytoalexins discovered have novel structures, even if they can often be related to
known natural products.
    The literature on phytoalexins has been admirably reviewed up
to 1982 in the book edited by Bailey and Mansfield (5). In this
paper, emphasis will be given to work carried out since then. Most

0097-6156/86/0296-0022$06.00/0

current research is concerned with the elicitation of phytoalexin synthesis, a process which can occur within a few hours of the appropriate stimulation. The elicitation mechanism has been well reviewed elsewhere (6) and will not be considered further here. Rather, I want to consider three related aspects of phytoalexins which are not so widely studied at present:

1) phytoalexin synthesis as a mechanism of natural disease resistance;

2) the distribution of this mechanism in higher plants;

and 3) the range of chemical structures encountered.

General aspects

Before proceeding further, it will be useful to briefly summarize our present knowledge of phytoalexin induction in plants. The following eight points may be made.

1) The known phytoalexins now number at least 200 chemical structures, over half of which have been characterized from a single family, the Leguminosae. A significant proportion of these structures are uniquely produced in this way and are not otherwise known as natural products.

2) Most phytoalexins can be classified as either masked phenolics (isoflavonoids, stilbenes, phenanthrenes, benzfurans) or as terpenoids. Fatty acid-derived polyacetylenes have been recorded in at least four plant families and nitrogen-containing phytoalexins have been described recently from the Caryophyllaceae and Gramineae.

3) Phytoalexins have been characterized from about 20 plant families, ranging from monocots to dicots and from trees to herbs (Table 1).

4) In any given plant, several substances are likely to be produced de novo. Occasionally, there may be a complex response: as many as 25 isoflavonoids have been isolated from the lima bean following abiotic induction (Table 2). Although some of the isoflavones detected are probably just biosynthetic intermediates, most accumulate in some quantity and have measurable fungitoxicities (7). Although usually related structures are found, chemically unrelated phytoalexins can be isolated in the response of particular plants.

5) Although fungitoxicity is the most characteristic biological property, phytoalexins may be antibacterial as well. Additionally, phytoalexins of legumes have been shown to be toxic to insects, while those from the sweet potato are poisonous to vertebrates (5).

6) Although usually bioassayed via their inhibition of fungal growth, phytoalexins can kill fungal cells. The principal site of action is probably the fungal plasmalemma but other sites are likely as well (5).

7) A variety of plant tissues besides the leaf will respond to induction: the hypocotyl, the stem, the germinating seed and woody tissue in trees. Elicitation of isoflavonoid and terpenoid phytoalexins has also been regularly achieved in plant cell cultures.

8) Elicitation is usually most effective with a fungus which is non-pathoenic on the plant being tested. Abiotic inducers, e.g. copper salt solutions, are also used and recently the enzyme pectinase has been shown to be particularly effective in causing phyoalexins to accumulate in fruit tissues (8).

Table 1. Chemical variation in phytoalexins within the flowering plants

Family	Genus	Chemical type	Structural example
Monocotyledons			
Amaryllidaceae	Narcissus	Flavan	7-Hydroxyflavan
Costaceae	Costus	Pterocarpan	Glyceollin II
Gramineae	Avena	Benzoxazin-4-one	Avenalumin I
	Oryza	Diterpene	Momilactone A
Orchidaceae	Orchis	Phenanthrene	Orchinol
Dicotyledons			
Caryophyllaceae	Dianthus	Benzoxazin-4-one	Dianthalexin
Chenopodiaceae	Beta	Isoflavone	Betavulgarin
Compositae	Carthamus	Polyacetylene	Safynol
	Helianthus	Coumarin	Scopoletin
	Lactuca	Sesquiterpene lactone	Costunolide
Convolvulaceae	Ipomoea	Furanoterpene	Ipomeamarone
Euphorbiaceae	Ricinus	Diterpene	Casbene
Leguminosae	Pisum, etc.	Isoflavonoid	
		Isoflavone	Wighteone
		Isoflavanone	Kievitone
		Pterocarpan	Medicarpin
		Isoflavan	Vestitol
	Lathyrus	Chromone	Lathodoratin
	Arachis	Stilbene	Resveratrol
	Vigna	Benzofuran	Vignafuran
	Vicia	Furanoacetylene	Wyerone
Linaceae	Linum	Phenylpropanoid	Coniferyl alcohol
Malvaceae	Gossypium	Naphthaldehyde	Gossypol
		Naphthaldehyde	Vergosin
Moraceae	Morus	Benzofuran	Muracin-C
		Stilbene	Oxyresveratrol
	Broussonetia	Flavan	7-Hydroxy-4'-methoxy flavan
Rosaceae	Eriobotrya	Biphenyl	Aucuparin
	Malus	Phenolic acid	Benzoic acid
	Pyrus	Dibenzofuran	γ-Pyrufuran
Solanaceae	Lycopersicon	Polyacetylene	Falcarinol
	Solanum, etc.	Sesquiterpene	Rishitin
Tiliaceae	Tilia	Sesquiterpene	7-Hydroxycala-menene
Ulmaceae	Ulmus	Terpenequinones	Mansonone A
Umbelliferae	Daucus	Polyacetylene	Falcarinol
		Dihydroiso-coumarin	6-Methoxy-mellein
	Pastinaca	Furanocoumarin	Xanthotoxin
Verbenaceae	Avicennia	Quinonoid	Naphthafuranone
Vitaceae	Vitis	Stilbene oligomers	α-Viniferin

The Phytoalexins of Legumes

The family most intensively investigated for phytoalexins is the
Leguminosae. Well over 500 species have now been screened at
Reading, using the drop-diffusate technique (9, 10). This is con-
venient in that it provides a phytoalexin solution, free from other
plant constituents. The disadvantage is that only milligram amounts
are available for identification. However, modern methods of
separation and spectral analysis usually operate successfully on
this scale.

The main point to observe is that virtually all species tested
gave a positive response. There is thus little doubt that phytoalexin synthesis is a universal attribute in this family. Even in
the genus Lupinus, which has been shown to have considerable numbers
of antifungal isopentenylisoflavones on the plant surface (11, 12),
phytoalexin induction could be detected (10).

The results of screening legumes have often proved of systematic interest. For example, in the tribe Vicieae where over 60 species were examined, a major dichotomy emerged (Table 3). The genera
Lathyrus and Pisum were united by the synthesis of pisatin and
related pterocarpans. By contrast, members of Vicia and Lens uniformly accumulate furanoacetylenes, such as wyerone (13), a compound
first characterized by Wain and his coworkers from broad bean in
1968 (14).

Some species in the Vicieae produce additional compounds to the
main types. We found, for example, that the sweet pea, Lathyrus
odoratus, proved to be almost unique in the tribe in synthesizing
two 3-ethylchromones. Such chromones had not been found before in
nature. Furthermore, one of them, 5,7-dihydroxy-3-ethylchromone,
was significantly more fungitoxic than pisatin (15). Last year,
Fuchs and his co-workers at Wageningen (16) reported further phytoalexins in the sweet pea, two α-hydroxydihydrochalcones, odoratol
and methylodoratol.

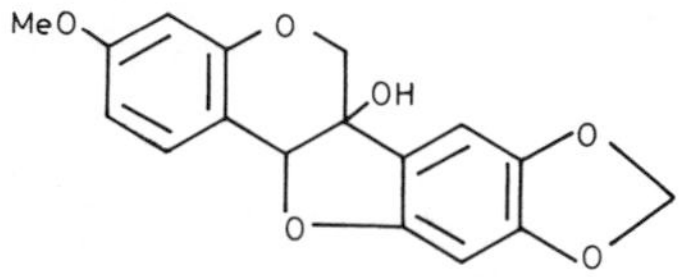

pisatin

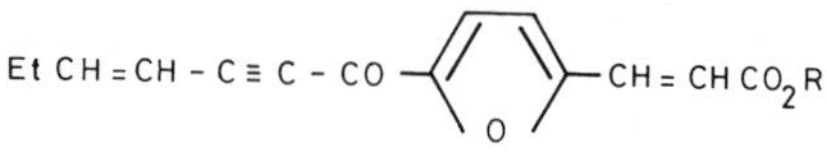

Et CH=CH - C ≡ C - CO ─ CH = CH CO$_2$R

wyerone   R = Me
wyerone acid   R = H
wyerone epoxide   C$_{11}$-C$_{12}$O

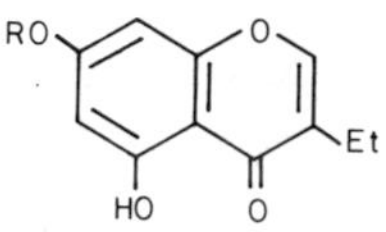

chromones   R = H and Me

Table 2. Inducible Isoflavonoids of Phaseolus lunatus

Class	Phytoalexin	Yield (μg/g fr wt x 10^{-3})
Isoflavone	Daidzein	--
	8-Hydroxygenistein	3.6
	Genistein	2.3
	8,2'-Dihydroxygenistein	5.8
	2'-Hydroxydaidzein	2.3
	2'-Hydroxygenistein	5.1
	2'-Methoxygenistein	5.1
	2,3-Dehydrokievitone	0.9
	2,3-Dehydrokievitol	0.8
	Luteone	0.2
	Cyclo-2,3-dehydrokievitone hydrate	2.1
	Isoferreirin	1.1
Isoflavanones	5-Deoxykievitone	2.3
	Kievitone	1.3
	5-Deoxykievitol	33.0
	Kievitol	1.9
	Kievitone hydrate	9.3
	3'-(γ,γ-Dimethylallyl) kievitone	9.7
	1",2"-Dehydrocyclokievitone	5.6
	Cyclokievitone hydrate	0.5
Pterocarpans	Phaseollidin	4.7
	4-(γ,γ-Dimethylallyl) phaseollidin	0.4
	2,10-Di(γ,γ-Dimethylallyl) glycinol	0.5
Coumestans	Coumestrol	1.2
	Psoralidin	--

Table 3. Phytoalexin differences within the tribe Vicieae

Genus	Number of species studied	Phytoalexin class	Compounds identified
Vicia	27)	Furanoacetylene	Wyerone and wyerone
Lens	2)		epoxide widespread
Pisum	2)	6a-Hydroxypterocarpan	Pisatin in all Pisum and
Lathyrus	31)		in 29 of 31 Lathyrus spp. representing 10 sections
Cicer	1	Pterocarpan	Medicarpin and maackiain

Data mainly from 13. Other compounds besides those mentioned were found in individual species. Traces of medicarpin have been found in Vicia faba, but the major response is furanoacetylene production. Lathyrus odoratus yields chromones and α-hydroxy dihydrochalcones as well (15, 16).

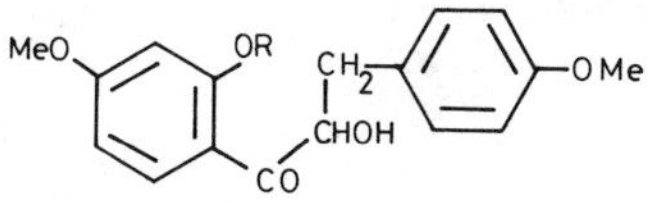

odoratol R=H
methylodoratol R=Me

Even the well known <u>Pisum</u> system originally studied by
Cruickshank and Perrin (4)has given new results. Pueppke and Van
Etten in 1976 (17) found additional phytoalexins were formed, later
in the interaction after pisatin was induced. And last year,
Carlson and Dolphin (18) reported production of cinnamylphenols and
a 2'-methoxychalcone in copper chloride-treated pea plants.

Thus, further probing of phytoalexin induction in a given species may well reveal the presence of new structures, not apparent in
the pioneering investigations. Biosynthetic investigations of phytoalexin synthesis in legumes (19) have also been useful in pro-
viding new insights into the induction process.

Phytoalexins in Other Dicotyledonous Families

At one time, it appeared that the same kind of phytoalexin was
likely to be encountered throughout a given plant family. However,
it is now clear that this is too simplistic and, as more species are
screened within families already investigated, unexpected chemical
types reveal themselves. For example, in the Solanaceae, a family
which has been investigated in detail in several laboratories, the
main type of phytoalexin is sesquiterpene-based. Over 20 structures
are known; the bicyclic sesquiterpene alcohol rishitin in typical as
is the spiro-compound solavetivone. These sesquiterpenes have been
found as phytoalexins in five genera: <u>Capsicum</u>, <u>Datura</u>, <u>Lycoper-
sicon</u>, <u>Nicotiana</u> and <u>Solanum</u> (Table 4). In the two latter genera,
species surveys have indicated that sesquiterpene synthesis is a
common response throughout (5). Yet recent examination of tomato
fruit and leaf tissue has shown that rishitin is accompanied by
three polyacetylenes (20, 21). One of the latter is falcarinol, a
C_{17} hydrocarbon first reported as a phytoalexin in the unrelated
carrot.

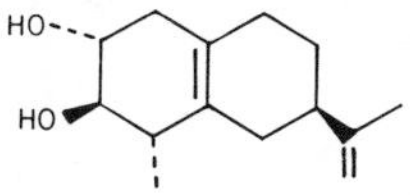

rishitin

$$CH_3(CH_2)_6\ CH=CHCHOH(C\equiv C)_2\ CHOH\ CH=CH_2$$

falcarindiol

$$CH_3(CH_2)_6CH=CH\ CH_2(C\equiv C)_2\ CHOH\ CH=CH_2$$

falcarinol

$$CH_3(CH_2)_5CHOH\ CH=CH\ CHOH\ (C\equiv C)_2\ H$$

Table 4. Sesquiterpenoid Phytoalexins of the Solanaceae

Phytoalexin*	Solanum (Potato)	Nicotiana (Tobacco)	Lycopersicon (Tomato)	Capsicum (Pepper)	Solanum (Eggplant)	Datura (Jimsonweed)
Rishitin	+	+	+	-	-	-
Lubimin	+	+	-	-	+	+
Phytuberin	+	+	-	-	-	-
Phytuberol	+	+	-	-	-	-
Solavetivone	+	+	-	-	-	-
Capsidiol	-	+	-	+	-	+
Glutinosone	-	+	-	-	-	-

* Other structures also produced in Potato (1), tomato (3), eggplant (4) and jimsonweed (2) which are species-specific

Whether acetylenes will be uncovered in other solanaceous plants remains to be seen. Tobacco has now been so thoroughly investigated that it is unlikely that acetylenes have yet to be discovered in its phytoalexin response.

Returning to the falcarinol production in carrot, it is still not clear whether this is typical of the family to which the carrot belongs, namely the Umbelliferae. Even in the carrot, the more characteristic fungitoxin is an isocoumarin, 6-methoxymellein (22). Yet another structural type – represented by the furanocoumarin xanthotoxin – has been characterized in infected parsnip tissues (23). Our own preliminary studies of several other umbellifer species suggest that furanocoumarin synthesis is the more typical response, but further work is needed to establish this.

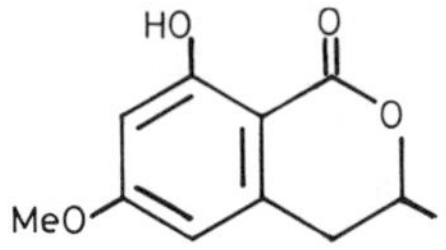

6 - methoxymellein

$$CH_3(CH_2)_6 \ CH = CH \ CH_2(C \equiv C)_2 \ CHOHCH = CH_2$$

falcarinol

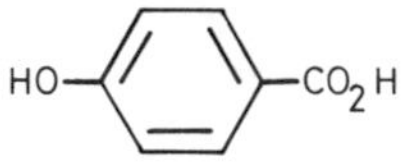

p - hydroxybenzoic acid

Yet another family producing polyacetylenes is the Compositae: two acetylenics safynol and dehydrosafynol are formed in the diseased safflower (24). Our own experiments at Reading indicate that acetylenes are formed in other members. This is however a very large family and other responses have been detected as well. Two sesquiterpene lactones have been encountered as phytoalexins in *Lactuca* (25) while coumarins are reported as such in the sunflower *Helianthus annuus* (26).

One speculative view is that acetylenes are a relatively primitive class of phytoalexin to be formed and we may expect other reports of these substances, probably co-occurring with more 'biogenetically advanced' phytoalexin types.

In the first years of phytoalexin research, most attention was given to crop plants, which are mostly herbs, and it was not clear whether trees responded in the same way. We now know that several trees are capable of phytoalexin accumulation.

A most detailed study by Japanese scientists of the mulberry tree has indicated a considerable range of benzfurans and other phenolics being formed in cortex and phloem tissues (27). Also, quinonoid phytoalexins, the mansonones, have been detected in infected elm trees (28). Furthermore, four methoxysubstituted dibenzofurans

(α- β- and γ-pyrufuran and cotonofuran) have been reported in two
rosaceous trees, the pear (29) and cotoneaster (30), while the
diphenyl, aucuparin, was detected after infection of a third rosaceous species, the loquat (31). Finally, a phenolic sesquiterpene
7-hydroxycalamenene has been characterized in the lime tree (32) and
several naphthofuranones from the grey mangrove (33). Trees thus
respond in various ways to fungal infection. Even the most primitive of all trees, Ginkgo biloba, has been shown to produce phytoalexins, although as yet the compounds formed have yet to be
properly characterized (34).

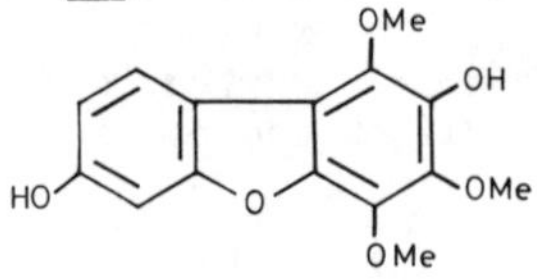

γ-pyrufuran

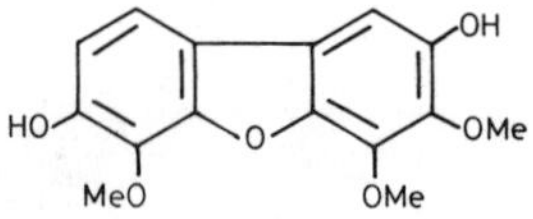

cotonefuran

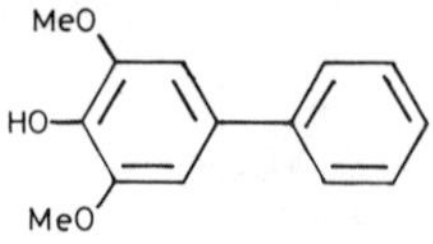

aucuparin

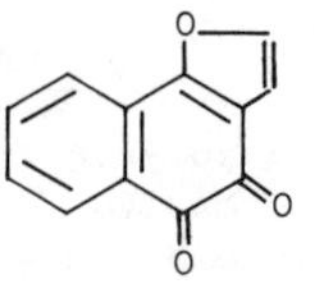

naphtho[1,2-b]furan-4,5-dione

Phytoalexins of Monocotyledonous Families

In spite of the enormous agricultural importance of the grasses,
relatively little is known of phytoalexin production in the
Gramineae. The monocotyledons have generally been neglected and
there are only records of phytoalexins being identified in about
four of these families. In chemical terms, there are some
surprising parallels with phytoalexin synthesis in the dicotyledons.
For example, the isoflavonoids glyceollins II and III first identified in soybean (Leguminosae), have unexpectedly appeared in
leaves of Costus speciosus (Costaceae) infected by Drechslera
longirostrata (35). Again, hydroxyflavans appear as phytoalexins in
dicots in the paper mulberry tree (27) and in monocots in daffodil
bulbs (36). Nevertheless, the best known monocot phytoalexins - the

hydroxyphenthrenes of orchid tissues – have yet to be reported in a
dicot source.

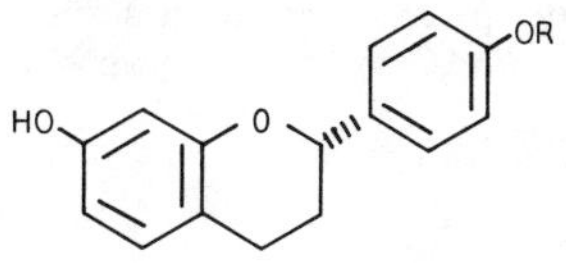

7, 4´-dihydroxyflavan   R = H

7-hydroxy - 4´- methoxyflavan R = Me

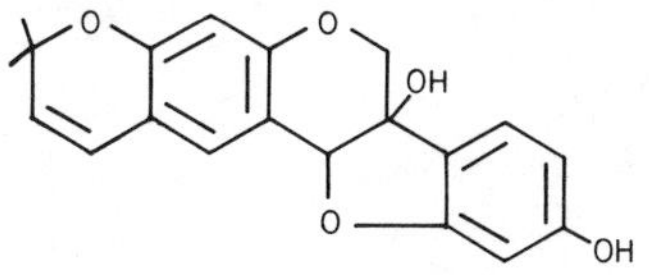

glyceollin II

Returning to the grasses, we now have positive identifications
of phytoalexins in oats and rice. In oats, avenalumins I, II and
III have been isolated from leaves infected with the crown rust
fungus. These three compounds are benzoxazin-4-ones substituted in
the 2-position with hydroxycinnamyl residues (37). Again there is a
parallel with phytoalexins of dicotyledons. A simple 2-phenyl
substituted benzoxazin-4-one dianthalexin was isolated from infected
carnation leaves by French scientists in 1983 (38).

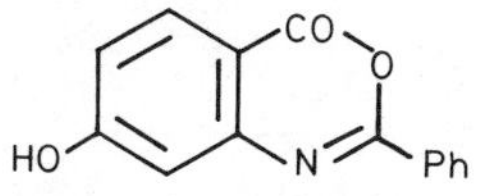

dianthalexin

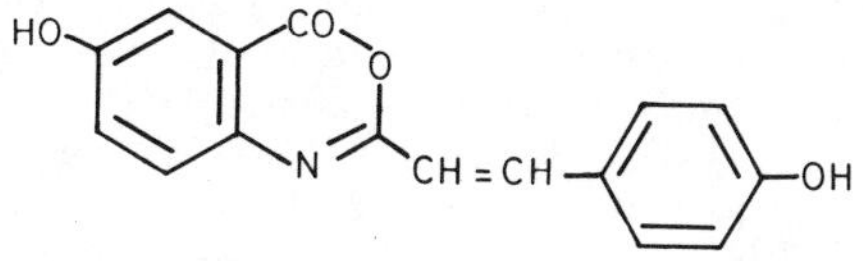

avenalumin 1

In the rice plant infected with the rice blast disease fungus,
two diterpene phytoalexins, momilactones A and B, were recorded in
1981 (39). These two substances incidentally also occur constitu-
tively in the husks of the same plant. More recently in 1984, three
further diterpene phytoalexins have been described oryzalexins A, B
and C (40). Structurally, these are sandaracopimaradiene derivati-

ves closely related to the two momilactones. Undoubtedly, the phytoalexin response of rice plants deserves further investigation and may well yield more compounds of this type in the future.

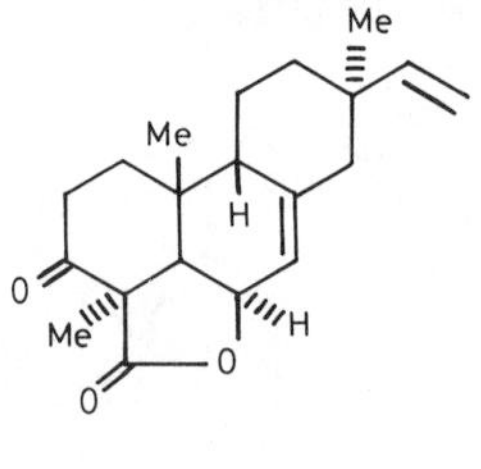

momilactone A

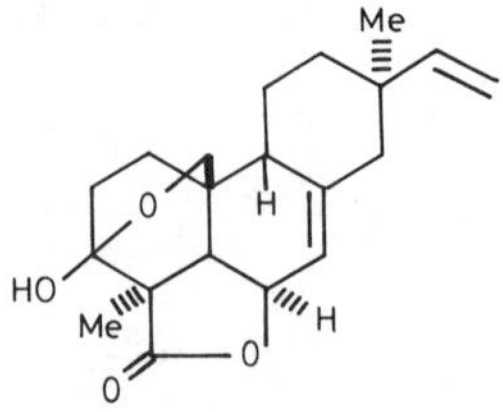

momilactone B

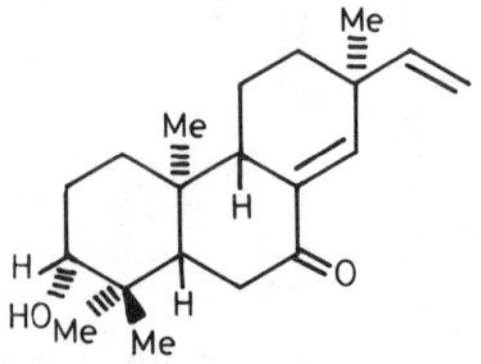

oryzalexin A

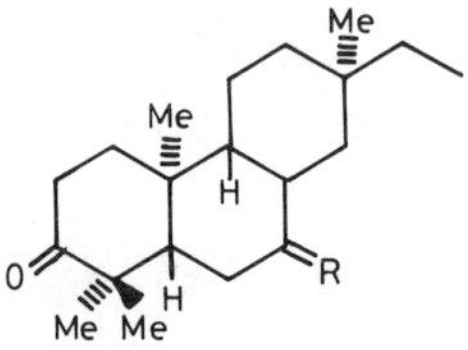

oryzalexin B   R = OH H

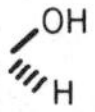

oryzalexin C   R = O

Conclusion

The results to date indicate that phytoalexin synthesis is a common mechanism of disease resistance in higher plants. Whether all plants will eventually yield positive results in this bioassay remains to be seen. It is our experience that there are still technical difficulties in establishing phytoalexin production in many plants and further development of appropriate methodology is essential. The separation of the phytoalexin response from the more general stress response of plants is also not easy and this needs to be considered in future experimental design.

Phytoalexin production per se is not absolute proof that it is an effective disease resistance mechanism in a given plant. It is at present difficult to evaluate the contribution of this antifungal barrier, along with other protective mechanism, to the disease immunity most plant species enjoy. The failure of phytoalexins to halt the progress of most pathogenic fungi that attack crop plants is not necessarily indicative of their ineffectiveness in nature. These are just the fungi which have developed, through coevolution with their chosen host plants, the biochemical machinery to detoxify the phytoalexins as soon as they are produced. Whether phytoalexin synthesis can turn the balance from susceptibility to resistance is a more interesting question to pose and there are some positive answers in certain host-plant relationships (41). These experiments open up the possibility of manipulating the phytoalexin trigger, possibly by application of synthetic chemicals which has the same effect as natural elicitors, so that it is able to prevent infection

from becoming established in a given crop plant. When resistant
varieties of a plant are bred, possibly what the plant breeder is
doing is selecting genetic ability to produce effective phytoalexins. An understanding of the biochemistry of phytoalexins may
lead to the ability to select and transfer this genetic ability from
resistant to susceptible varieties.

To the pesticide scientist, phytoalexin research should still
be of interest in spite of the fact that nothing commercially useful
has yet emerged. There are ample indications that a considerable
number of novel fungitoxins remain to be uncovered in plants. These
compounds are distinctive from constitutive antifungal substances;
often a different biosynthetic pathway is induced from those
involved in normal secondary metabolism. Relatively few plants have
yet been analysed in depth so that we still have a long way to go to
understand fully this important disease resistance mechanism in the
plant kingdom as a whole.

That only small amounts of phytoalexin are available usually
from laboratory experiments would seem to limit the opportunities
for pesticide screening. However, it is usually possible to scale
up procedures to yield up to gram amounts. The placing of a
suitable fungal spore suspension in opened pea pods, for example,
will produce a solution from which pisatin will crystallize (4).
The successful elicitation of phytoalexins in plant cell culture
also opens the way to large scale production. Plant culture may
incidentally provide additional metabolites to those formed in
intact plants (42).

Even if the phytoalexins so far isolated have little commercial
utility, it is still possible that useful substances may yet appear.
But the more exciting possibility is that consideration of the chemical structures of natural phytoalexins and of their modes of
action against fungal infections may provide clues for the development of synthetic pesticides. The complexity of the chemical structures of the natural phytoalexins may make them uneconomical to
manufacture but a comparison should be made with the synthetic
pyrethroid insecticides. The natural pyrethrums have complex chemical structures but simpler compounds, economical to manufacture,
have been developed on the basis of the structures of natural
pyrethrums and many of these have much more desirable properties for
use in agriculture than the natural substances. There would seem to
be no reason why simpler compounds based on the structures of
natural phytoalexins should not provide synthetic fungicides as
important and useful as the synthetic pyrethroids. This is a future
challenge for the synthetic organic chemists in this area.

Literature Cited

1. Schultz, J. C.; Baldwin, I. T. *Science* 1982, 217, 149.
2. Ward, H. M. *Ann. Bot.* 1905, 19, 1.
3. Muller, K. O.; Borger, H. *Arb. biol. Reichsanst. Land-u.
    Forstw. Berlin-Dahlem* 1941, 23, 189.
4. Cruickshank, I. A. M.; Perrin, D. R. *Nature* 1960, 187, 799.
5. Bailey, J. A.; Mansfield, J. W. (eds.) *"Phytoalexins"*;
    Blackie, Glasgow, 1982.
6. Darrill, A. G.; Albersheim, P. *Ann. Rev. Pl. Physiol.* 1984,
    35, 243.

7.   O'Neill, M. J.; Adesanya, S. A.; Roberts, M. F. Z. Natur-
     forsch. 1983, 38c, 693.
8.   Watson, D. G.; Brooks, C. J. W. Physiological Plant Pathology
     1984, 24, 331.
9.   Harborne, J. B.; Ingham, J. L. In "Biochemical Aspects of
     Plant and Animal Coevolution" (J. B. Harborne, ed.) Academic
     Press, London, 1978, p. 343.
10. Ingham, J. L. Fortschritte d. Chem. org. Naturst. 1983, 43, 1.
11. Harborne, J. B.; Ingham, J. L.; King, L.; Payne, N. Phyto-
     chemistry 1976, 15, 1485.
12. Ingham, J. L.; Tahara, S.; Harborne, J. B. Z. Naturforsch.
     1983, 38c, 194.
13. Robeson, D. J.; Harborne, J. B. Phytochemistry 1980, 19, 2359.
14. Fawcett, C. H.; Spencer, D. M.; Wain, R. L.; Fallis, A. G.;
     Jones, E. R. H.; Le Quan, M.; Page, C. B.; Thaller, V.;
     Shubrook, D. C.; Witham, P. M. J. Chem. Soc. 1968, 2455.
15. Robeson, D. J.; Ingham, J. L.; Harborne, J. B. Phytochemistry
     1980, 19, 2171.
16. Fuchs, A.; De Vries, F. W., Landheer, C. A.; Van Veldhuizen, A.
     Phytochemistry 1984, 23, 2199.
17. Pueppke, S. G.; VanEtten, H. D. J. Chem. Soc., Perkins Trans.
     I. 1975, 946.
18. Carlson, R. E.; Dolphin, D. H. Phytochemistry 1982, 21, 1733.
19. Banks, S. W.; Dewick, P. M. Phytochemistry 1983, 22, 1591.
20. De Wit, P. J. G. M.; Kodde, E. Physiol. Plant Path. 1981, 18,
     143.
21. Engersma, D. M.; Overeem, C. Neth. J. Plant Path. 1981, 87,
     69.
22. Condon, P.; Kuc, J. Phytopathology 1960, 50, 267.
23. Johnson, C.; Brannon, D. R.; Kuc, J. Phytochemistry 1973, 12,
     2961.
24. Allen, E. H.; Thomas, C. A. Phytopathology 1972, 62, 471.
25. Takasugi, M.; Okinaka, S.; Katsui, N.; Masamune, T.; Shirata,
     A.; Ohuchi, M. J. C. S. Chem. Comm. 1985, 621.
26. Tal, B.; Robeson, D. J. Phytochemistry 1985, 24, in press.
27. Takasugi M.; Kumagai, Y.; Nagao, S. Chemistry Letters 1980,
     1459.
28. Dumas, M. T.; Strunz, G. M.; Hubber, M.; Jung, R. S.
     Experientia 1983, 39, 1089.
29. Kemp, M. S.; Burden, R. S. J. Chem. Soc. Perkin Trans. I.
     1984, 1441.
30. Burden, R. S.; Kemp, M. S.; Wiltshire, C. W.; Owen, J. D. J.
     Chem. Soc. Perkin Trans. I. 1984, 1445.
31. Watanabe, K.; Ishiguri, Y.; Nonaka, F.; Morita, A. Agric.
     Biol. Chem. 1982, 46, 567.
32. Burden, R. S.; Kemp, M. S. Phytochemistry 1983, 22, 1039.
33. Sutton, D. C.; Gillan, F. T.; Susie, M. Phytochemistry 1985,
     24, in press.
34. Ingham, J. L. Bot. Rev. 1972, 38, 343.
35. Kumar, S.; Shukla, R. S.; Singh, K. P.; Paxton, J. D.; Husain,
     A. Phytopathology 1984, 74, 1349.
36. Coxon, D. T.; O'Neill, T. M.; Mansfield, J. W.; Porter, A. E.
     A. Phytochemistry 1980, 19, 889.

37. Mayama, S.; Tani, T.; Ueno, T.; Hirobayashi, K.; Nakashima, T.;
     Fukami, H.; Mizuno, Y.; Irie, H. Tetrahedron Letters 1981, 22,
     2103.
38. Bouillant, M. L.; Favre-Bonvin, J.; Ricci. P. Tetrahedron
     Letters 1983, 24, 51.
39. Cartwright, D.; Langcake, P.; Pryce, R. J.; Leworthy, D. P.;
     Ride, J. P. Nature 1977, 267, 511.
40. Kono, Y.; Takeuchi, S.; Kodama, O.; Akatsuka, T. Agr. Biol.
     Chem. Tokyo 1984, 48, 253.
41. Keen, N. T. in "Plant Disease Control, Resistance and
     Susceptibility" (R.C. Staples and G. H. Toenniessen, eds.)
     Wiley, New York, 1981, p. 155.
42. Watson, D. G.; Rycroft, D. S.; Freer, I. M.; Brooks, C. J. W.
     Phytochemistry 1985, 24, in press.

RECEIVED October 3, 1985

4

Biochemistry of Cotton (*Gossypium*) Resistance to Pathogens

Alois A. Bell, Marshall E. Mace, and Robert D. Stipanovic

National Cotton Pathology Research Laboratory, USDA, Agricultural Research Service, College Station, TX 77841

Pathogens that cause diseases of cotton can be divided into three classes based on cellular reactions in susceptible and resistant plants as shown in Table 1.

Table 1. Classification of pathogens based on cellular reactions in susceptible and resistant cotton plants.

Class of Pathogen	Cellular Reaction of Plant	
	Susceptible	Resistant
Biotroph	Remains alive adjacent to pathogen	Rapid confined death adjacent to pathogen
Transient biotroph	Delayed death adjacent to pathogen	Rapid confined death adjacent to pathogen
Necrotroph	Extensive death well beyond pathogen	Rapid confined death adjacent to pathogen

In all cases resistance involves the rapid death of plant cells at the infection site. This reaction is often referred to as a hypersensitive reaction (HR) when a biotroph or transient biotroph is the pathogen. We prefer to avoid this terminology, because it implies that there is some unusual sensitivity in the resistance response to these organisms; in fact, as shown in Table 1, rapid cell death is the normal resistance response to nearly all pathogens that attempt to invade a potential host. We prefer to call this response necrogenic resistance (NR), contrasting it with resistance to nonpathogens or avirulent pathogens, which does not involve plant cell death. NR is an active process involving greatly accelerated metabolism and accumulation of various natural products (antibiotics, enzyme denaturants, etc.) that contribute to the resistance reaction (1, 2).

The differences in classes of pathogens are in the mode of virulence (ability to overcome resistance). The biotrophic pathogen either actively suppresses or does not induce the NR response in the susceptible host. In this situation, chemicals or enzymes associated with NR either are not formed or occur in very low concentrations. Examples of biotrophic pathogens attacking cotton are the rust fungi, root-knot and reniforme nematodes, and viruses.

Transient biotrophic pathogens also suppress NR in the susceptible host, but only long enough to allow the pathogen to "outrun" the resistance response, while the pathogen colonizes the host. Thus, chemicals associated with NR are formed quickly and occur at the highest concentrations in resistant hosts 12-72 hrs after infection. In susceptible hosts the slow response allows more extensive colonization and, as a result, more elicitation of resistance spatially after 1 to 2 weeks. Thus, more accumulation of response chemicals may eventually occur in the susceptible plant. The Fusarium and Verticillium wilt fungi and the bacterial pathogen Xanthomonas campestris pv. malvacearum are examples of important transient biotrophic pathogens of cotton.

Nectrotrophic pathogens rapidly elicit NR in either susceptible or resistant plants, and in some cases secrete small molecular weight phytotoxins that diffuse through susceptible tissues and elicit NR even at considerable distances from the pathogen. In such cases the resistance response may be triggered and dissipated before contact of resistance chemicals is made with the pathogen. Virulence of necrotrophs also may be associated with very fast growth rates of the pathogen or with the ability to inactivate compounds formed during NR. The necrotrophic pathogens are also opportunistic, attacking juvenile or old tissues that have weak defense responses. Examples of necrotrophic fungal pathogens in cotton are the seedling pathogen Rhizoctonia solani, the boll rot and seed pathogens Diplodia gossypina and Aspergillus niger, and the leaf spot pathogen Alternaria macrospora. The biochemistry of cotton resistance to each of the three classes of pathogens will be discussed in later sections.

Resistance is a relative term used to describe the ability of a plant to prevent, restrict, or retard the penetration and development of pathogens in host tissue. The biochemical events involved with resistance are invariably invoked in "susceptible" as well as in resistant plants in response to transient biotrophs and necrotrophs. Thus, from a biochemical perspective susceptible plants also resist infection, but the speed or intensity of the response is inadequate, and the pathogen is able to progressively colonize the plant. Therefore, the degree of resistance depends on the speed and intensity of the defensive response of the host relative to the speed of infection by the pathogen (3). Repeated sampling at various intervals after inoculation is necessary to properly monitor biochemical mechanisms of resistance.

Some components of resistance are present in healthy plants and therefore are part of a constitutive defense; other components are induced and formed following contact between the pathogen (or its products) and the host and therefore are parts of an active defense. The total defense is the sum of constitutive and active components. Many groups of chemicals function in both constitutive and active defense. Therefore, I will discuss the components of defense based on chemical similarities, noting how they serve constitutive and active roles.

Chemical Components of Defense

Terpenoid Aldehydes. The tribe Gossypieae is distinguished from other malvaceous plants by the presence of lysigenous glands, called pigment glands. In seed of Gossypium species, the glands predominantly contain the terpenoid aldehyde gossypol (Fig; 1). Gossypol is responsible for the toxicity of cottonseed to a variety of herbivores, including most monogastric mammals and various insects (4). In foliage of different Gossypium species the pigment glands contain gossypol and 15 additional terpenoid aldehydes (5) and over 20 volatile terpenes (6). In addition, high concentrations of anthocyanins occur in the epithelial cells surrounding the gland cavity (7). The structures of these compounds and their importance in resistance to insects is discussed in another paper by Stipanovic and Williams in this volume. Pigment glands in resistant plants can rupture and release their contents into intercellular spaces in response to microbial pathogens (8). Thus, the glands might be a constitutive defense component against certain microorganisms.

Gossypol and its methyl ethers (Fig. 1) also are formed in the epidermis and root hairs of the developing seedling root (9) and later are infused throughout the periderm of the root bark of older plants. Terpenoids also are exuded by roots, and exudation is increased by microbial infections (10, 11). Only the root tip is devoid of terpenoids, which may explain why this is the only area of the root penetrated by root-knot nematode or fungal wilt pathogens.

Terpenoid aldehydes and their naphthofuran precursors are formed by various cotton tissues as phytoalexins (ie., antibiotic chemicals synthesized in response to microbial infection). The primary phytoalexins of this type in most Gossypium species are hemigossypol (HG) and its precursor desoxyhemigossypol (dHG), their 3-methyl ethers (MHG and dMHG), and gossypol and its mono- (MG) and dimethyl ether (DMG) shown in Fig. 1. Two Gossypium species also form raimondal (R) and presumably desoxyraimondal (dR) as minor phytoalexins. These phytoalexins, except for R and dR, are made predominantly by tissues devoid of chloroplasts; R and dR are formed in green tissues. Terpenoid aldehydes and precursors are the major phytoalexins in the root, young hypocotyl, stele, and endocarp of the boll (fruit). The ratios of HG and MHG to G, MG, and DMG vary considerably among tissues. For example, G is barely detectable in infected xylem vessels (12), but is equally prevalent with HG in nematode infected roots (13). The relative percentages probably are determined by peroxidase activity in the specific tissue; peroxidase dimerizes HG to G (14). Relative concentrations of naphthofurans, which are probably the most toxic of these phytoalexins (15), are greatest in infected tissues at 12-48 hrs after inoculation (16). These compounds are readily autoxidized to the sequiterpenoid aldehydes (HG, MHG, R) and consequently are transient; yet toxic levels may persist for several weeks (15). The addition of the 3-methyl group apparently stabilizes the naphthofuran against oxidation, allowing accumulation of greater concentrations of dMHG than of dHG.

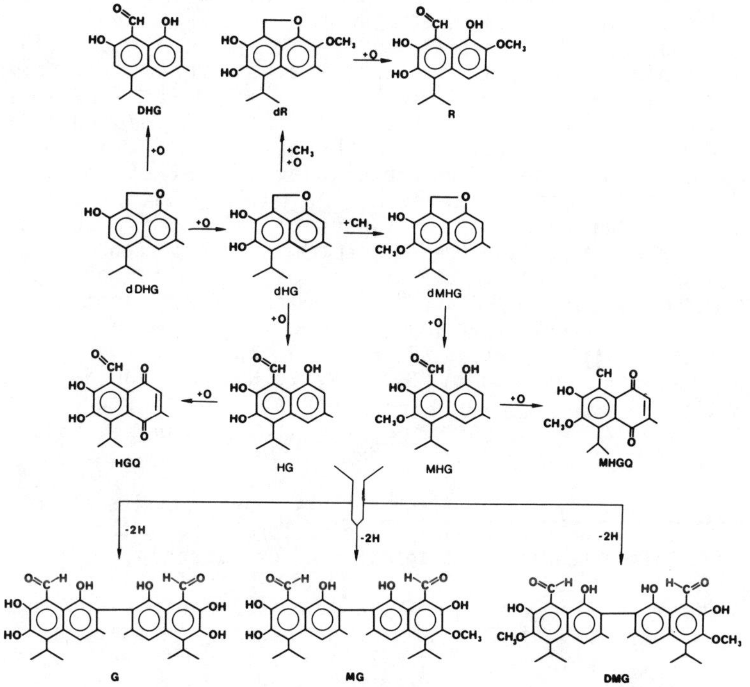

Fig. 1. Structure of naphthofuran and terpenoid aldehyde phytoalexins formed by gossypium species. See text for identifications.

The major phytoalexins formed by green tissues (cotyledons, leaves, bracts, and the epicarp of bolls) are the cadalenes and lacinilines shown in Fig. 2 (17, 18). About 75% of these compounds are methylated in the cultivated G. hirsutum (19), whereas methylation is either greatly restricted or absent in the other cultivated cottons (G. arboreum, G. herbaceum and G. barbadense). These compounds have been demonstrated in palisade parenchyma cells (20), but probably also occur in other parenchyma cells with abundant chloroplasts. Recent biosynthetic studies of the terpenoid aldehyde (21) and cadalene (22) phytoalexins indicate that the direction of terpenoid biosynthesis to one or the other of these groups of compounds may be determined by the folding patterns of farnesyl pyrophosphate during cyclization, since different folding patterns have been found for the biosynthesis of hemigossypol and 2,7-dihydroxycadalene. The subcellular localization of these reactions is uncertain. However, the chloroplasts are suspect, because their differentiation apparently is associated with the changes in the flow of the terpenoid pathway from terpenoid aldehydes to cadalenes. Two other reactions, the methoxylation of dHG to dR and the oxidation of HG or MHG to hemigossypolone (HGQ) or hemigossypolone methyl ether (MHGQ) are associated with chloroplast differentiation (5). That is, R, HGQ, and MHGQ are found in green tissue, but are replaced by HG, MHG, G, MG, and DMG in nongreen tissue of roots, stele, and seed embryos. Plastids are known to contain the polyphenoloxidases (23), which probably catalyze the oxidation of HG to HGQ and dHG to dR.

Condensed Proanthocyanidins (Tannins). A representative structure of the condensed proanthocyanidins (tannin) formed in cotton is shown in Fig. 3. Cotton tannins contain mixtures of catechin, epicatechin, gallocatechin and epigallocatechin moieties in the polymers. The free catechins also occur, but in lower concentrations than the tannins. The ratio of catechin to gallocatechin moieties in the polymer generally varies from 1:1 to 1:4 in different cultivated cottons. Very little gallocatechin occurs in some of the wild species.

The tannins are potent protein denaturants and consequently act as enzyme inhibitors, antisporulants, and mild antibiotics (24, 25, 26). The tannins also are intermediates in the synthesis of the dark brown melanoid pigments found in diseased tissues and seed coats. The melanins restrict water flow through seed coats (27) and probably restrict water loss from wound sites. The tannins generally are located in vacuoles within cells of the endodermis and hypodermis and in scattered parenchyma cells of other tissues. Tannin concentrations as high as 20-40% (dry wt.) occur in buds, young leaves, young bolls, and young bracts (28). Infection may cause additional parenchyma cells to synthesize tannins, especially in the xylem ray cells of the stele (16). Consequently tannin levels may increase from 0.5-1.0 to 5-10% following vascular invasion by microbial pathogens. The tannins probably are synthesized by endoplasmic reticulum, collecting in very small vacuoles that latter coalesce to form the few vacuoles found in mature tannin containing cells (29). Cells that contain high levels of tannins also show high

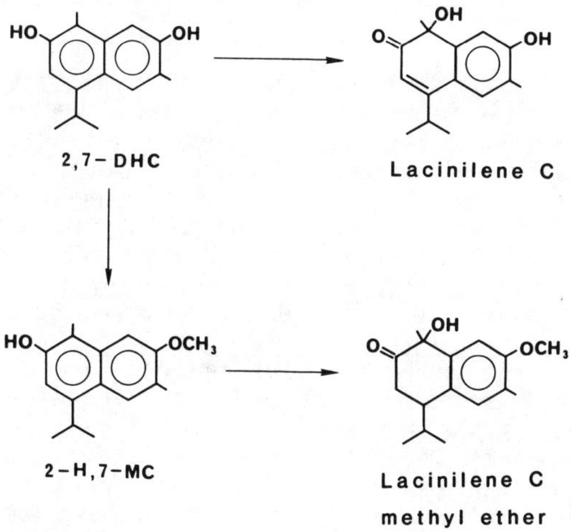

Fig. 2.   Structures and biochemical relationships of 2,7-dihydroxycadalene, 2-hydroxy-7-methoxycadalene, lacinilene C, and lacinilene C methyl ether.

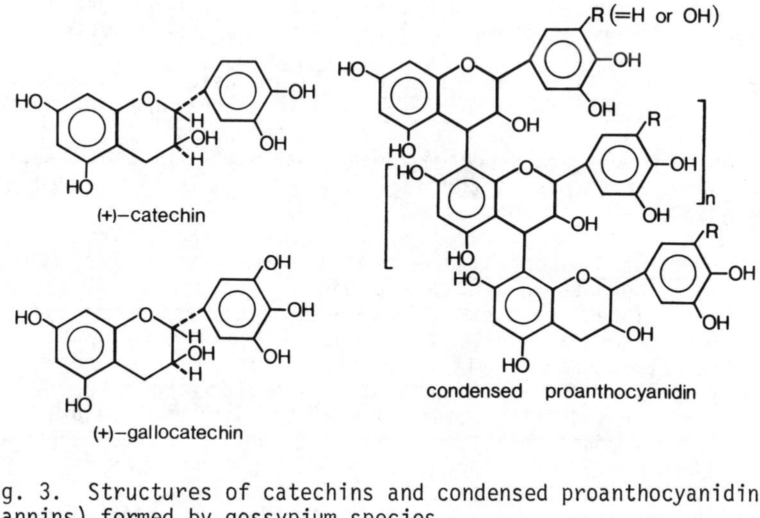

Fig. 3.   Structures of catechins and condensed proanthocyanidins (tannins) formed by gossypium species.

levels of peroxidase and polyphenoloxidase activity, indicating that
these enzymes may be involved in tannin synthesis and oxidation to
melanin (23,30)

Cutin, Suberin, and Wax. Cells exposed to air or the soil generally
are covered by a cutin or suberin layer coated with wax. These
polymeric sheets provide a physical barrier to penetration by some
microorganisms and they also prevent leakage of water and nutrients
from the cytoplasm to the plant surface. These structures have been
shown to be important for resistance to microbial infection in other
plants (31), but they have received limited attention in cotton.
    Two recent studies (32,33) have partially elucidated the
structure of cutin and suberin in cotton. Cotton suberin is unusual
in that it contains predominantly C-22 fatty acids
(22-hydroxydocosanoic and 1,22-docosanedioic) in the polymer.
Quantitation of these acids might be useful to determine changes in
suberin content during NR responses. Cutin in cotton is composed
mostly of C16 and some C18 fatty acid derivatives, and is very
similar to that in other plants.
    Cotton strains and species with green fiber have lamellar layers
(up to 26) of suberin and wax deposited alternately with cellulose
during formation of secondary walls in the epidermal cells, including
fiber cells, of the seed coat. Seeds of cotton with green lint are
less permeable to water than those with white lint, indicating that
suberin may be involved in regulating water uptake by seeds.

Lignin. The importance of lignin in resistance to microbial
infection has been demonstrated in other crops (1). However, few
studies have been concerned with the role of lignin in resistance in
cotton, even though more than 20% of the dry matter of a mature stem
is lignin. The ratio of syringyl to guaiacyl units in cotton lignin
varies with age (34), species (35,36), and infection by
microorganisms (37). G. barbadense cultivars have a predominance of
syringyl units, whereas G. hirsutum cultivars contain mostly guaiacyl
units. Both internal condensation within the polymer and percentages
of methoxyl groups are reported to increase with age and infection.
About 75% of the bonds in the lignin polymer are alkyl-aryl ether
bonds and 25% are C-C bonds (38).

Other Possible Components. Lipids that might be important as
defensive components in cotton are the cyclopropenoid fatty acids
(39), hydroxylated unsaturated fatty acids (40), and other oxidation
products of unsaturated fatty acids (41). Similar fatty acids are
antibiotic and act as self defensive compounds in rice (42).
Cottonseed oil contains only about 1% each of cyclopropenoid and
hydroxy unsaturated fatty acids. However, oil from immature seeds,
radicles (very young roots) and root tips contain high concentrations
(up to 28%) of cyclopropenoid acids (39), suggesting a possible
protective role for these fatty acids in these tissues.
    Cotton tissues also may contain 1-3% flavonoid glucosides, most
of which contain ortho-dihydroxyphenolic groups (5). Additional
quantities of flavonoid glucosides are formed in response to
infection. These compounds are important in constitutive defense
against insects, but their importance in the resistance of cotton to

pathogens is unknown. Mixtures of several ortho-dihydroxyphenols with peroxidase are antibiotic against Xanthomonas phaseoli, a bacterial pathogen (43). Thus, mixtures of cotton flavonoids and peroxidase also might have antibiotic activity.

Resistance to Classes of Pathogens.

Biotrophs. Necrogenic resistance in cotton has been demonstrated against several biotrophic pathogens: the southwestern cotton rust fungus, the tropical rust fungus, blue disease virus, anthocyanosis virus, reniform nematode and root knot nematode (44, 45). However, only resistance to root-knot has been studied in detail.

The root knot nematode, Meloidogyne incognita, includes 4 races, of which only races 3 and 4 are able to establish a compatible (susceptible) feeding relationship with the host (46). In the susceptible host the nematode physically penetrates through the root cortex and establishes a feeding site in the pericycle, called the giant cell. The animal becomes sedentary, so that maintenance of the giant cell is essential for feeding. The metabolically hyperactive giant cell contains numerous nuclei and exhibits extensive protein synthesis. Active defense reactions are not apparent around the giant cell.

In resistant hosts, pericycle cells near the head of the sedentary animal become necrotic, and a marked reduction in egg laying is seen (both in numbers of egg masses and in eggs/mass). Thus, reductions in eggs (or egg masses) per gram of root is normally used as evidence of resistance (47).

Concentrations of constitutive terpenoids in the root epidermis of cotton are unrelated to differences in resistance. But concentrations of terpenoid aldehydes formed in the vicinity of the pericycle, near the head of the animal, act as phytoalexins and are closely correlated with levels of resistance (11, 48). Little or no phytoalexin is formed in the pericycle of susceptible cultivars. Mixtures of terpenoid phytoalexins are more toxic to the nematode than gossypol alone, and mixtures containing methylated terpenoid phytoalexins (from G. hirsutum) are more toxic than those that contain only nonmethylated phytoalexins (from G. arboreum) (49). Thus, the structure of terpenoid phytoalexins also is important for resistance to root knot nematode.

Veech (50,51) looked for possible relationships between tannins and resistance. In general, the highest tannin levels are found in susceptible cultivars, and tannin levels decrease following infection. The greatest decrease (ca. 35%) occurs in the most resistant cultivar. These data indicate that tannin synthesis is not part of the active defense against root knot nematode. The disappearance of tannin might be associated with its conversion to melanin or other oxidation products, but the significance of such changes for resistance to nematodes is unknown.

Transient Biotrophs (Fungi). Resistance to the wilt fungi, Fusarium oxysporum f. sp. vasinfectum and Verticillium dahliae, has been studied extensively. Details of these studies have been reviewed (16, 52)

Verticillium penetrates cotton roots directly just back of the

root tip where all tissues are devoid of terpenoids and tannin
(9,53). The fungus penetrates through the cortex eventually entering
the xylem vessel, where conidia are formed in the xylem stream.
These conidia are carried to xylem vessel end walls where they
germinate and penetrate through perforation plates into the next
vessel element or through vessel side walls into adjoining vessels.
New conidia are again formed and the cycle is repeated until the
entire vascular system of the plant is invaded. As terminal stem or
leaf tissues die, the fungus invades cells surrounding the vessels
and forms microsclerotia which survive in soil to repeat the disease
cycle. Fusarium has a similar infection pattern, except that it
initially invades primarily through wounds or nematode feeding areas,
and chlamydospores are formed in the moribund tissue.

The containment of both fungi in xylem vessels depends on the
rapid sequential formation of tyloses which seal the vessel above the
invading fungus and terpenoid aldehyde and naphthofuran phytoalexins
which kill or stop the growth of the fungus (54,55). Mace et al.
(15) did a detailed study of the toxicity of phytoalexins to V.
dahliae. The phytoalexin dHG was the only one sufficiently toxic and
water soluble to kill the fungus. The other phytoalexins required
DMSO or surfactant to solubilize toxic concentrations. It is not
known whether natural surfactants are exuded with phytoalexins in
cotton.

The importance of terpenoid phytoalexins in resistance to wilts
was further demonstrated in studies of temperature effects on
resistance and in studies of induced resistance. Increase in
temperature from 25 to 30°C causes a marked increase in resistance.
This temperature change also slows the rate of sporulation of the
fungus and increases the rate of phytoalexin formation by cotton
(56). Likewise, treatments that induce resistance also induce
phytoalexin synthesis (57). Phytoalexin synthesis, therefore, is
also important to explain environmental effects on disease
resistance.

Histochemical studies have shown that the phytoalexins are
formed in paravascular parenchyma cells, and are exuded into xylem
vessels (58). Since some parenchyma cells contain predominately dHG,
it is possible that the naphthofurans alone are synthesized in the
parenchyma cells and exuded into the vessels and intercellular
spaces, where they readily oxidize to terpenoid aldehydes.

Correlative studies have indicated that constitutive terpenoid
aldehydes in the root epidermis may also contribute to resistance.
Epidermal cells in healthy roots of cultivars resistant to V. dahliae
tended to form terpenoids at earlier stages of development, but did
not necessarily contain greater concentrations when mature (59).
Likewise gossypol concentrations in 2-cm root tips from 7-day-old
seedlings of 18 cultivars were significantly (r = 0.59 in field and
0.77 in greenhouse) correlated with resistance to Fusarium (11).
These correlations might result from the restriction of penetration
by gossypol in the root epidermis. However, the greenhouse
inoculation techniques bypassed the epidermis. Thus, the high levels
of gossypol in young epidermal tissues of resistant plants probably
reflect the greater potential to form terpenoid aldehyde phytoalexins
in various tissues of these plants.

Tannin synthesis is also activated more rapidly and occurs more intensely in resistant than susceptible cultivars in response to V. dahliae (16). Additional tannin synthesis also occurs in young infected leaves but not old infected leaves (24). The tannins in stele tissue are formed throughout the xylem ray cells which surround xylem vessels (60). Therefore the tannins may be important in restricting fungal growth to xylem vessels. Tannins have potent antisporulant and weak antibiotic activity against V. dahliae (24).

Constitutive tannins also may contribute to resistance. The resistant G. barbadense cultivars have higher levels of tannins, especially in older leaves, than do susceptible G. hirsutum cultivars (16). Also, the G. barbadense tannins appear to be more astringent, based on their antisporulant activities. Endodermal cells several millimeters back of the root tip also contain large amounts of tannin (53). Many attempted penetrations of the stele are aborted at the endodermis, which forms a protective ring around the vascular cylinder.

Transient Biotrophs (Bacteria). The bacterium Xanthomonas campestris pv. malvacearum enters cotton leaves or stems through water-congested stomata or wounds and initiates growth on the surface of parenchyma cells (61). In susceptible cultivars water-soaking of tissues first occurs, and tissues later became chlorotic and eventually necrotic (blighted). Bacterial populations of ca 10^8 - 10^9/g are often reached in susceptible host tissue. In resistant cultivars the initial multiplication of bacteria is similar to that in susceptible cultivars; a rapid NR then occurs giving rise to small clusters of dark brown cells. Bacterial populations are curtailed at about 10^6 - 10^7/g of leaf tissue (17, 62). An antibiotic environment apparently develops concurrent with the cessation of bacterial multiplication (63).

Resistant, compared to susceptible, cultivars exhibit more rapid synthesis of terpenoid aldehyde phytoalexins in xylem vessels (16), and cadalene and lacinilene phytoalexins in cotyledonary tissue (17), in response to X. campestris infection. Of the various compounds tested, 2,7-dihydroxycadalene, followed by lacinilene C, had the greatest antibiotic activity against the bacterium. Lacinilene C exists as two optical isomers, which occur in different proportions in two different resistant cultivars. The (+)-lacinilene C is more than three times as toxic as (-)-lacinilene C (17). The toxicity of terpenoid aldehydes and their precursors have not been tested against the bacterium.

Pigment glands in resistant cultivars are reported to rupture more readily in resistant than susceptible cultivars (8), but the toxicity of compounds released from the glands has not been tested against X. campestris . Thus, the role of constitutive terpenoids in resistance is uncertain.

No detailed studies of tannins in bacterial infected cotton have been made. However, the formation of dark brown pigments is a part of the necrogenic response to X. campestris. Brown pigments in cotton have been shown to originate from oxidation of tannins (27, 64), most of which have been newly synthesized in diseased tissue (65). Venere (62) found a marked increase in extractable peroxidase activity in resistant but not susceptible tissue in response to X. campestris, and mixtures of peroxidase and catechin (a monomeric unit of tannin) were bactericidal. Thus, he concluded that oxidation of catechin by peroxidase may be a factor in restricting growth of the bacterial pathogen in blight resistant cotton. Similar, or more striking, results might have been obtained if tannin had been used as the substrate. Cotton leaves and cotyledons normally contain 5-10% (dry wt.) tannin (5); thus, appreciable amounts would be available if released from vacuoles into the intercellular spaces.

Necrotrophs. Resistance to necrotrophic pathogens in cotton has been most studied with the sore shin disease of seedlings caused by Rhizoctonia solani. Only minor differences occur in the resistance of cultivars or species to the pathogen, but tissues of different ages show marked differences in resistance. The seed and young hypocotyl and radicle of all cultivars are susceptible to infection, whereas hypocotyls two or more weeks old are resistant. The change from the susceptible to resistant state occurs gradually over a period of about 10 days. Polygalacturonase secreted by the pathogen is thought to be important in the killing and maceration of host tissue.

Hunter et al. (25, 26, 66) have examined the roles of terpenoids, catechins, and tannins in the resistance of old and young hypocotyls to R. solani. Differences in resistance of 5- or 6-day-old compared to 12- or 14-day-old seedlings are most closely correlated with increases in the constitutive compounds, especially tannins and catechins, that occur with aging. Infection elicits rapid synthesis of terpenoid phytoalexins beginning prior to 24 hr and continuing until 48 hr after inoculation of either age hypocotyl. Beyond 48 hr there is a decline in naphthofurans and no further increase in sesquiterpenoid aldehydes in the degenerating susceptible tissue, while all phytoalexins continue to increases in concentration in the resistant tissues (66). Infection of hypocotyls also increases the exudation of terpenoids from the root epidermis (10). The quantities of tannins formed in response to infection are greater in 12-day-old than 5-day-old plants at 24 hrs, but not 48 hrs, after infection (25). In all cases the total of constitutive plus actively synthesized tannins or terpenoids are greatest in the resistant tissue.

The effects of catechins and tannins on mycelial growth and on polygalacturonase (PG) production and activity of R. solani also were determined (25,26). Catechin at concentrations found in 14-day-old hypocotyls causes a 50-90% inhibition of growth of three isolates of the fungus, whereas concentrations found in 5-day-old hypocotyls had little or no effect on growth. A highly virulent fungal isolate is less sensitive to catechin than are two moderately virulent isolates. Catechin also inhibits PG production of the moderately viruluent isolates, but not of the virulent isolate. Catechin mixed with hypocotyl enzymes or added to culture filtrates forms oxidation products that inactivated PG. Hunter (26) concluded that catechin and polyphenols (tannins) are important in age-related resistance, and act by inhibiting both fungal growth and PG activity.

The role of tannins and terpenoids in the resistance of cottonseed to necrotrophic pathogens also has been investigated. Cottonseed does not form phytoalexins until the seed moisture level reaches 20% or higher (67); at this level germination has already begun. Thus, phytoalexins are not involved in resistance to pathogens, such as Aspergillus spp. that can attack seed at moisture levels as low as 12%. The constitutive terpenoids in the lysigenous glands of the embryo also do not offer any appreciable protection, because glandless seed is no more susceptible to rot than glandular seed. Halloin (68) concluded that the major protection of the seed comes from components that restrict rates of water imbibition, thus maintaining low seed moisture levels during alternate periods of wetting and drying in the field.

When seeds removed from the boll a few days before boll opening are ripened under nitrogen, the tannins in the pigment layer are not converted to melanins, and seed coats are highly permeable to water (64). Therefore, tannins in immature seed coats may give protection against microbial invasion, whereas oxidation products of tannin (melanin) in mature seed may be important for restricting water uptake, thereby preventing seed deterioration. Recent observations (32,33) on cottons with green fiber indicate that suberins also may be important in regulating water uptake and therefore contribute resistance.

Cotton bolls, like hypocotyls, show age-related changes in resistance to necrotrophic pathogens that cause boll rots. Middle aged bolls generally show NR responses to most potential pathogens. As bolls age, however, resistance is gradually lost, and bolls in their last week of development often become susceptible. The biochemical bases for this loss of resistance has been studied by several investigators.

The endocarp tissue (inner tissue) of the boll synthesizes terpenoid aldehyde phytoalexins in response to infection or chemical stress (69,70), whereas the epicarp (outer green tissue) synthesizes cadalenes and lacinilenes (71). In each case synthesis is initiated rapidly regardless of cultivar or pathogenic species, although higher concentrations of both groups of phytoalexins eventually accumulate in the middle aged bolls. Most boll rot fungi are relatively insensitive to the terpenoid aldehydes (69). Some, such as Fusarium moniliforme, can completely destroy the antibiotic aldehyde function of gossypol in less than 24 hours, when the compound is added to 2-day-old cultures at mM concentrations (unpublished data). The toxicity of lacinilenes and cadalenes have not been studied. Without demonstrated toxicity or substantial concentrations, phytoalexins alone do not appear to account for the developmental change in resistance (70,72).

Several observations indicate that differences in constitutive defense are important for age-related differences in boll resistance to necrotrophic fungal pathogens. Cuticle thickness and quantity of waxes and cutin acids are greatest in intermediate aged bolls and decrease with the decline in resistance (72). Various extracts of the cuticle also are inhibitory to the spore germination of numerous boll rotting pathogens (73). Condensed tannin concentrations also relate to resistance. The boll wall may contain over 20% tannin, most of which is in the epicarp. Puncture inoculations into the epicarp with Diplodia gossypina give a typical NR, and infection is stopped, but inoculation into the pericarp or through the wall into the fiber allows a susceptible reaction (70). The fiber is devoid of tannins and is highly susceptible to attack by many necrotrophic fungi that gain access to it through wounds regardless of boll age.

Intense browning occurs around puncture inoculations of middle aged bolls, indicating oxidation of tannins to melanins and probably to toxic intermediates. An increase in peroxidase and the appearance of new peroxidase isozymes has been reported in infected resistant bolls (74). However, others (75) have been unable to confirm this observation. Both groups of investigators, however, found high levels of peroxidase in healthy tissue, so this enzyme still may be responsible for oxidizing tannins.

Summary of Resistance-Class Interactions. Studies of resistance to various classes of pathogens indicate that active resistance mechanisms, especially phytoalexin synthesis, are very effective against biotrophs, and constitutive defenses are relatively unimportant. Constitutive defenses complement active defenses to give the highest level of resistance against transient biotrophs. In contrast to resistance to biotrophs, active resistance mechanisms are not very effective against necrotrophs, unless a high level of constitutive defense is present. Thus, levels of constitutive defense components largely determine the level of resistance to necrotrophs.

For active defenses to be effective they must be implemented quickly. A delay of 24-48 hr in the "recognition" of a transient biotroph can be the difference between a susceptible and resistant reaction as has been shown for the fungal wilt diseases. Therefore, it is important to review what is known about recognition and compounds that elicit active defense responses.

Recognition and Elicitors

Chemicals or treatments that effect active defense responses are referred to as elicitors, and may be either biotic or abiotic in origin. Abiotic elicitors of terpenoid aldehyde and tannin synthesis in cotton include chilling injury, ultraviolet irradiation, cupric ions, and various pesticides (69, 76, 77, 78, 79). Biotic elicitors include live and dead fungal spores and bacterial cells (16). Various heterogeneous polymers obtained from fungal and bacterial cell walls also act as elicitors. These include an extracellular polysaccharide (EPS) from X. campestris (16), a lipoproteinpolysaccharide (LPS) from V. dahliae (80) and a glycoprotein from V. dahliae (81). Of the nonliving biotic elicitors, only dead cells of V. dahliae and EPS from X. campestris elicit greater active defense responses in resistant cultivars than in susceptible cultivars (16). Thus, specific recognition may involve some component of the outer cell wall of the microbial pathogen.

Studies with other plant-pathogen systems have shown that oligomers from chitosan and β-1,3-glucans can be potent elicitors (82, 83). Both chitin and the glucans are common constituents of fungal cell walls and are cleaved by chitinases and β-1,3-glucanases that occur in plants; concentrations of these enzymes increase in many plants soon after infection (1). A specific oligomer size (usually 5-8 subunits) of chitosan or β-1,3-glucans is required for appreciable activity. It is possible that such sizes are cleaved from LPS, EPS and dead cells of fungal and bacterial pathogens to activate defense responses in cotton. A β-N-acetylglucosaminidase has been demonstrated in cotton tissue (84).

Certain pathogenic organisms produce antigens that cross-react with antibodies produced against cotton antigens (85). These common antigens usually are not found in nonpathogenic organisms. A common antigen isolated from the Fusarium wilt pathogen of cotton was identified as a protein-carbohydrate complex similar to that in cell walls (86). Thus, virulent pathogens may synthesize outer cell walls that are similar to their hosts. This might prevent breakdown by host enzymes and consequent release of oligomeric elicitors.

Pectate oligomers also may act as elicitors and can be formed by the action of fungal or host pectinase enzymes on pectin. Pectinase enzymes cause browning and necrosis in cotton (87, 88), and release peroxidase from cotton cell walls (89). Disturbances of cell wall structure in cottons infected with X. campestris can be shown as early as 2 hrs after inoculation in resistant, but not susceptible, cultivars (90). Pectinase also can be demonstrated 1-3 days earlier in the infected resistant compared to susceptible tissue (91). Thus, pectate fragments may also serve as elicitors of NR in cotton, but this remains to be proven.

Elicitation of active defense responses does not require an exogenous elicitor, since physical treatments such as UV irradiation and chilling can be very potent elicitors. Spontaneous active defense reactions also occur extensively in interspecific hybrids of Gossypium undergoing genetic lethal reactions (92). Thus, alterations in membranes or activation of chemical reactions in cell walls or on membranes may ultimately control active defense.

Symptoms and Active Defense

Symptoms of disease are considered to be due to physical and chemical damage inflicted on the plant by the pathogen and its metabolites. However, there is increasing evidence that cotton disease symptoms are due, at least partially, to the active defense response which, if extensive enough, may kill the plant. Several observations support this possibility.

Browning, a classical symptom of plant disease, has been recognized as being caused by the oxidation of polyphenols. However, this often was considered a passive process in which phenols were allowed to mix with oxidases released from compartmentalization during breakdown of tissues by pathogens. Studies of Verticillium wilt, bacterial blight, and boll rots indicate that phenols and even the oxidases involved in browning may be formed as part of the active defense (62, 65, 88, 91).

The most convincing correlation between symptoms and active defense responses have come from our studies of genetic lethal reactions in three different interspecific hybrids of Gossypium: G. hirsutum x G. gossypioides; G. hirsutum x G. davidsonii; and G. hirsutum x G. arboreum var sanguineum. In each hybrid the lethal reaction is due to a single gene in the wild species interacting with one or two genes in G. hirsutum (45, 93, 94). Lethality can be prevented by first introducing interspecific compatibility (C) genes into G. hirsutum (45, 95) from G. barbadense. A different C gene(s) is needed for each diploid species. The genetics of interspecific incompatibility is very similar to the gene-for-gene systems purported to control plant resistance to biotrophic pathogens.

We (unpublished) have found that plants undergoing the lethal reaction show most of the symptoms considered classical for plant disease. Dark brown necrotic lesions develop on leaves, stems, and embryos of young seedlings of G. hirsutum x G. davidsonii, and plants collapse and die as young seedlings. The G. hirsutum x G. gossypioides hybrids develop symptoms that are very similar to symptoms of fungal wilt disease: epinasty, chlorosis, and loss of turgor in leaves; and necrotic lesions, tyloses, hypertrophy, and hyperplasia in the stele and cambium. Plants die after 10-12 true leaves are formed, but young seedlings give no evidence of the impending lethal response. The G. hirsutum x G. arboreum hybrids also develop symptoms similar to wilt diseases, but the onset of the lethal reaction often is not apparent until early flowering, and in winter months in the greenhouse lethality may not occur until some viable seeds are produced.

We have examined these lethal reactions biochemically (93, unpublished). In all cases spontaneous defense reactions (phytoalexin synthesis, tannin synthesis and tannin oxidation) occur concurrent with the onset of symptoms. The quantities of terpenoid aldehydes and tannins found in tissues of severely affected plants are usually very similar to those found in severely diseased plants dying from microbial infections. Thus, it is possible that many of the disease symptoms seen in cotton are due to the toxic terpenoids and tannins formed in response to infections.

Applications

The usual stated purpose for studying the biochemistry of resistance is to develop information that may be used by geneticists to improve disease resistance in cultivars. Therefore, we will speculate briefly on approaches that might be taken. Each approach depends on the class of pathogen involved.

For biotrophs NR is very effective in containing the pathogen. When NR is available in the Gossypium species of concern, a system of hybridization and backcrossing can be used to transfer it. The use of several genes that condition NR in combination gives stable resistance, whereas resistance conditioned by single genes often succumbs to previously unknown strains of the pathogen. The development of bacterial blight resistance (96) and root-knot nematode resistance (97) in cotton are excellent examples of using multiple genes to condition NR. When no resistance is available in the cultivated species, NR might be located in a wild diploid cotton and transferred to the cultivated tetraploid, using various hybridization and chromosome doubling schemes (45). It also may be possible to create NR where none exists by making multiple-species hybrids. We have obtained high levels of resistance to Verticillium wilt and root-knot nematode from G. hirsutum x (G. arboreum x G. raimondii) hybrids, even though none of the individual species show NR. Another possible source of NR might be the genes conditioning interspecific incompatibility, since they do affect the spontaneous NR responses in hybrid plants. Each of these possibilities needs to be explored.

For transient biotrophs better sources of NR also are desirable, but it also may be possible to increase resistance by improving antibiotic efficiency. Genetic elimination of methylation in the cadalenes and lacinilenes should give more toxic phytoalexins without added energy cost to the plant. In contrast, adding more methylation to the terpenoid aldehyde pathway may stabilize the naphthofuran structure and allow greater accumulation of the naphthofuran relative to its aldehyde derivative. The net effect of this change is difficult to predict, because the very labile dHG is more toxic than dMHG and more water soluble, even though it will not accumulate to equally high concentrations.

Increasing tannin astringency is another desirable change. This should be possible by increasing the percentage of gallocatechin residues in the polymer and optimizing polymer size. Such changes may have adverse effects on the tonoplast, or even be responsible for lethal reactions. Therefore, transfer of additional genes to alter tonoplast structure may be necessary to accomodate the more astringent tannins.

Increased resistance to necrotrophs also could result from changes that increase antibiotic efficiency. Other approaches for these pathogens should aim at increasing constitutive defenses. Such things as higher tannin and lignin concentration, thicker cuticle structure, and increased suberization would be expected to improve resistance.

Literature Cited

1. Bell, A. A. Ann. Rev. Plant Physiol. 1981, 32, 21-81.
2. Bell, A. A. In "Vegetative Compatibility Responses in Plants"; Moore, R., Ed.; Baylor University Press: Waco, TX, 1983; pp. 47-69.
3. Bell, A. A. In "Plant Disease: An Advanced Treatise"; Horsfall, J. G.; Cowling, E., Eds.; Academic: New York, 1980; Vol. V, pp. 53-73.
4. Bell, A. A.; Stipanovic, R. D. Proc. Beltwide Cotton Prod. Res. Conf., 1977, pp. 244-58.
5. Bell, A. A. In "Cotton Physiology -- A Treatise"; Stewart, J. M.; Mauney, J. R., Eds.; In press.
6. Elzen, G. W.; Williams, H. J.; Bell, A. A.; Stipanovic, R. D.; Vinson, S. B. Proc. 9th Cotton Dust Res. Conf., 1985, pp. 47-8.
7. Chan, B. G.; Waiss, A. C., Jr. Proc. Beltwide Cotton Prod. Res. Conf., 1981, pp. 49-51.
8. Novacky, A. Plant Dis. Reptr. 1972, 56, 765-7.
9. Mace, M. E.; Bell, A. A.; Stipanovic, R. D. Phytopathology. 1974, 64, 1297-302.
10. Hunter, R. E.; Halloin, J. M.; Veech, J. A.; Carter, W. W. Plant and Soil. 1978, 50, 237-40.
11. Hedin, P. A.; Shepherd, R. L.; Kappelman, A. J., Jr. J. Agric. Food Chem. 1984, 32, 633-8.
12. Mace, M. E.; Bell, A. A.; Beckman, C. H. Can. J. Bot. 1976, 54, 2095-9.
13. Veech, J. A. Nematologica. 1978, 24, 81-7.

14. Veech, J. A.; Stipanovic, R. D.; Bell, A. A. J. C. S. Chem.
     Comm. 1976, p. 144-5.
15. Mace, M. E.; Stipanovic, R. D.; Bell, A. A. Physiol. Plant
     Path. 1985, 26, 209-18.
16. Bell, A. A.; Stipanovic, R. D. Mycopathologia. 1978, 65,
     91-106.
17. Essenberg, M.; Doherty, D. A.; Hamilton, B. K.; Henning, V. T.;
     Cover, E. C.; McFaul, S. J.; Johnson, W. M. Phytopathology.
     1982, 72, 1349-56.
18. Zeringue, H. J., Jr. Phytochem. 1984, 23, 2501-3.
19. Bell, A. A.; Stipanovic, R. D.; Greenblatt, G. A.; Williams, H.
     J.; Elzen, G. W. In "Cotton Dust: Advances in Occupational
     Health and Fiber Cleaning"; Montalvo, J. G., Ed.; (In press).
20. Essenberg, M.; Cover, E. C.; Pierce, M. L.; Richardson, P. E..
     Phytopathology. 1982, 72, 945.
21. Widmaier, R.; Howe, J.; Heinstein, P. Arch. Biochem. and
     Biophys. 1980, 200, 609-16.
22. Essenberg, M.; Stoessl, A.; Stothers, J. B. J. C. S. Chem.
     Comm. 1985, 556-7.
23. Mueller, W. C.; Beckman, C. H. Can. J. Bot. 1978, 56, 1579-87.
24. Howell, C. R.; Bell, A. A.; Stipanovic, R. D. Physiol. Plant
     Path. 1976, 8, 181-8.
25. Hunter, R. E. Physiol. Plant Path. 1974, 4, 151-9.
26. Hunter, R. E. Phytopathology. 1978, 68, 1032-6.
27. Halloin, J. M. New Phytol. 1982, 90, 651-7.
28. Chan, B. G. Proc. 9th Cotton Dust Res. Conf., 1985, pp. 49-52.
29. Meuller, W. C.; Beckman, C. H. Can. J. Bot. 1976, 54, 2074-82.
30. Veech, J. A. Phytopathology. 1976, 66, 1072-6.
31. Kolattukudy, P. E. Science. 1980, 208, 990-1000.
32. Ryser, U.; Holloway, P. J. Planta. 1985, 163, 151-63.
33. Yatsu, L. Y.; Espelie, K. E.; Kolattukudy, P. E. Plant Physiol.
     1983, 73, 521-4.
34. Veksler, N. A.; Smirnova, L. S.; Abduazimov, K. A. Khim. Prir.
     Soedin. (Tashk). 1977, 1, 100-7.
35. Fahmy, Y.; Mobarak, F.; Schweers, W. Cellulose Chem. Technol.
     1982, 16, 453-9.
36. Mirzakhmedova, M. K.; Smirnova, L. S.; Abduazimov, K. A. Chem.
     Nat. Comp. 1983, 19, 591-3.
37. Pulatov, B. K.; Abduazimov, K. A. Chem. Nat. Comp. 1983, 19,
     517-8.
38. Smirnova, L. S.; Abduazimov, K. A. Chem. Nat. Comp. 1978, 14,
     430-1.
39. Fisher, G. S.; Cherry, J. P. Lipids. 1983, 18, 589-94.
40. Stipanovic, R. D.; Donovan, J. C.; Bell, A. A.; Martin, F. W.
     J. Agric. Food Chem. 1984, 32, 809-10.
41. Vick, B. A.; Zimmerman, D. C. Plant Physiol. 1981, 67, 92-7.
42. Kato, T.; Yamaguchi, Y.; Hirano, T.; Yokoyama, T.; Uyehara, T.;
     Namai, T.; Yamanaka, S.; Harada, N. Chemistry Letters. 1984,
     pp. 409-12.
43. Urs, N. V. R. R.; Dunleavy, J. M. Phytopathology. 1975, 65,
     686-90.
44. Bell, A. A. In "Protection Practices in the USA and World";
     Kohel, R. J.; Lewis, C. F., Eds.; AM. AGRON. SOC. MONOGRAPH
     SERIES No. 24, American Agronomy Society: Madison, WI, 1983;
     pp. 288-309.

45. Bell, A. A. In "Phytochemical Adaptations to Stress";
    Timmermann, B. N.; Steelink, C.; Loewus, F. A., Eds.; RECENT
    ADVANCES IN PHYTOCHEMISTRY Vol. 18, Plenum Press: New York,
    1984, pp. 197-229.
46. Veech, J. A. In "Protection Practices in the USA and World";
    Kohel, R. J.; Lewis, C. F., Eds.; AM. AGRON. SOC. MONOGRAPH
    SERIES No. 24, American Agronomy Society: Madison, WI, 1983;
    pp. 309-30.
47. Veech, J. A. J. Nematol. 1982, 14, 2-9.
48. Veech, J. A.; McClure, M. A. J. Nematol. 1977, 9, 225-9.
49. Veech, J. A. J. Nematol. 1979, 11, 240-6.
50. Veech, J. A. Proc. Beltwide Cotton Prod. Res. Conf., 1977, p.
    37.
51. Veech, J. A. J. Nematol. 1979, 11, 316.
52. Bell, A. A.; Mace, M. E. Proc. Beltwide Cotton Prod. Res. Conf.
    1984, pp. 43-7.
53. Mace, M. E.; Howell, C. R. Can. J. Bot. 1974, 52, 2423-6.
54. Mace, M. E. Physiol. Plant Pathol. 1978, 12, 1-11.
55. Harrison, N. A.; Beckman, C. H. Physiol. Plant Pathol. 1982,
    21, 193-207.
56. Bell, A. A.; Presley, J. T. Phytopathology. 1969, 59, 1141-6.
57. Bell, A. A.; Presley, J. T. Phytopathology. 1969, 59, 1147-51.
58. Mace, M. E. New Phytol. 1983, 95, 115-9.
59. Bell, A. A. Phytopathology. 1969, 59, 1119-27.
60. Mace, M. E.; Bell, A. A.; Stipanovic, R. D. Physiol. Plant
    Pathol. 1978, 13, 143-9.

61. Essenberg, M. K.; Cason, E. T.; Hamilton, B.; Brinkerhoff, L.
    A.; Gholson, R. K.; Richardson, P. E. Phyiol. Plant Pathol.
    1979, 15, 53-68.
62. Venere, R. J. Plant Sci. Lett. 1980, 20, 47-56.
63. Essenberg, M.; Hamilton, B.; Cason, E. T.; Brinkerhoff, L. A.;
    Gholson, R. K.; Richardson, P. E. Physiol. Plant Pathol. 1979,
    15, 69-78.
64. Halloin, J. M. Proc. Beltwide Cotton Prod. Res. Conf., 1976, p.
    45.
65. Bell, A. A.; Simpson, M. E.; Marsh, P. B.; Howell, C. R. Proc.
    Beltwide Cotton Prod. Res. Conf., 1971, p. 84.
66. Hunter, R. E.; Halloin, J. M.; Veech, J. A.; Carter, W. W.
    Phytopathology. 1978, 68, 347-50.
67. Halloin, J. M.; Bell, A. A. J. Agric. Food Chem. 1979, 27,
    1407-9.
68. Halloin, J. M. Proc. Beltwide Cotton Prod. Res. Conf., 1984, p.
    28.
69. Bell, A. A. Phytopathology. 1967, 57, 759-64.
70. Mace, M. E.; Halloin, J. M. Proc. Beltwide Cotton Prod. Res.
    Conf., 1982, p. 46.
71. Halloin, J. M.; Greenblatt, G. A. Proc. Beltwide Cotton Prod.
    Res. Conf. 1982, p. 46.
72. Roberts, R. G.; Snow, J. P. Phytopathology. 1984, 74, 390-7.
73. Wang, S. C.; Pinckard, J. A. Phytopathology. 1973, 63, 315-9.
74. Wang, S. C.; Pinckard, J. A. Phytopathology. 1973, 63, 1181-5.
75. Wang, S. C.; Pinckard, J. A. Phytopathology. 1973, 63, 1095-9.
76. Mellon, J. E.; Lee, L. S. Plant Physiology. 1983, 72, 24.
77. Bell, A. A.; Christiansen, M. N. Phytopathology. 1968, 58,
    883.

78. Komives, T.; Casida, J. E. J. Agric. Food Chem. 1983, 31,
     751-5.
79. Parrott, W. L.; Lane, H. C. Proc. Beltwide Cotton Prod. Res.
     Conf. 1980, p. 88.
80. Grinstein, A.; Lisker, N.; Katan, J.; Eshel, Y. Physiol. Plant
     Pathol. 1984, 24, 347-56.
81. Zaki, A. I.; Keen, N. T.; Erwin, D. C. Phytopathology. 1972,
     62, 1402-6.
82. Heinstein, P. Planta Medica. 1980, 39, 196-7.
83. Keen, N. T.; Yoshikawa, M.; Want, M. C. Plant Physiology.
     1983, 71, 466-71.
84. Kendra, D. F.; Hadwiger, L. A. Experimental Mycology. 1984, 8,
     276-81.
85. Yi, C. K. Proc. Oklahoma Acad. Science. 1981, 61, 36-40.
86. Charudattan, R.; DeVay, J. E. Phytopathology. 1972, 62, 230-4.
87. Charudattan, R.; DeVay, J. E. Physiology Plant Pathol. 1981,
     18, 289-95.
88. Hooper, D. G.; Venere, R. J.; Brinkerhoff, L. A.; Gholson, R. K.
     Phytopathology. 1975, 65, 206-13.
89. Wang, S. C.; Pinckard, J. A. Phytopathology. 1972, 62, 460-5.
90. Strand, L. L.; Mussell, H. Phytopathology. 1975, 65, 830-1.
91. Cason, E. T.; Richardson, P. E.; Essenberg, M. K.; Brinkerhoff,
     W. M.; Johnson, W. M.; Venere, R. J. Phytopathology. 1978, 68,
     1015-21.
92. Venere, R. J.; Brinkerhoff, L. Proc. Amer. Phytopathol. Soc.
     1974, 1, 78-9.
93. Mace, M. E.; Bell, A. A. Can. J. Bot. 1981, 59, 951-5.
94. Gerstel, D. U. Genetics. 1954, 39, 628-39.
95. Lee, J. A. J. Heredity. 1981, 72, 299-300.
96. Brinkerhoff, L. A.; Verhalen, L. M.; Johnson, W. M.; Essenberg,
     M.; Richardson, P. E. Plant Disease. 1984, 68, 168-73.
97. Shepherd, R. L. Crop Science. 1983, 23, 999-1002.

RECEIVED October 8, 1985

Regulation of Accumulated Phytoalexin Levels in Ladino Clover Callus

David L. Gustine

U.S. Regional Pasture Research Laboratory, USDA, Agricultural Research Service, University Park, PA 16802

Thiol/disulfide ratios and the interaction of biological disulfides with plasma membrane proteins are postulated as mechanisms by which Ladino clover callus cells regulate biosynthesis and accumulation of the phytoalexin, medicarpin. Evidence is presented to show that organic mercurial sulfhydryl (SH) reagents and N-ethyl maleimide react with free sulfhydryl groups in the cell wall membrane of callus cells. Mersalyl, a non-penetrating SH reagent, was a weak elicitor of medicarpin production in callus, while other sulfhydryl reagents tested in the callus system elicited levels as high or higher than levels reported to be elicited in plants. Biological dithiols, cystine and oxidized glutathione, were also elicitors. Incorporation studies with ^{14}C-L-phenylalanine and ^{14}C-acetate showed that medicarpin accumulated in the callus cells because its rate of biosynthesis was increased. Results from these studies support the hypothesis that membrane constituents containing sulfhydryl groups are important factors in regulation of phytoalexin synthesis.

Plant tissue culture is an extremely useful tool for investigating biochemical and hormonal aspects of plant metabolism. Such systems have been applied to studies of host-pathogen interactions and secondary metabolism. My laboratory has been investigating the biochemical regulation of phytoalexin accumulation in host-pathogen interactions of cultured legumes.

Phytoalexins are generally recognized as important components of induced resistance in many plants. These metabolites are not found in healthy plant tissues or are present at very low concentrations. They accumulate as a result of interactions between plant cells and elicitor molecules. Fragments of fungal or plant cell walls can act as biological elicitors, apparently

formed by enzymatic hydrolysis of glucan or pectin polysaccharides
(1). Phytoalexins are effective antibiotics because they quickly
accumulate in high concentrations at the site of invasion (1).
Ample evidence now exists to demonstrate a direct role for
phytoalexins in preventing successful infection by fungi (2-3-4)
and bacteria (5).

Phytoalexins in Alfalfa and Clovers

At least 10 pterocarpanoid phytoalexins have been identified in
alfalfa and clover species. These antibiotic metabolites have
been elicited by fungi, bacteria, and a variety of abiotic
substances. Biosynthesis of pterocarpanoid phytoalexins begins
with phenylalanine (6) and follows the pathway outlined by Martin
and Dewick (7), depicted in Figure 1. Medicarpin, sativan,
isosativan, and vestitol are produced in most clovers and alfalfa
in response to biotic and abiotic elicitors. In clovers,
elicitors also induce accumulation of maackiain, isovestitol, and
arvensin (8).

The role of these phytoalexins in resistance to diseases in
alfalfa and clovers has not been fully resolved. Studies of
alfalfa and red clover showed that medicarpin or maackiain is
accumulated in response to pathogens, but data are lacking on the
concentration and localization of these metabolites at the
infection sites (9-10-11-12-13). Limited evidence suggests that
rates of synthesis and concentrations of medicarpin and/or sativan
in alfalfa leaf and stem tissues are related to resistance to
Verticillium albo-atrum (14-15-16). Thus it appears that
phytoalexins at least partly restrict growth of fungi in these
forage legumes. This is not unexpected since phytoalexins are
components of resistance in many plant/pathogen interactions
involving fungi, bacteria, nematodes, or viruses (17).

Phytoalexins in Legume Tissue Cultures

Tissue Cultures Respond to Elicitors. Recent investigations in my
laboratory have established that callus cultures of Canavalia
ensiformis (jackbean), Medicago sativa (alfalfa), and nine species
of Trifolium (clover) biosynthesize phytoalexins in response to
elicitors in essentially the same way that leaves, hypocotyls, or
seedlings do. When callus tissue cultures were exposed to spores
of Pithomyces chartarum (18) or sulfhydryl (SH) reagents (19-20),
medicarpin accumulated at concentrations higher than those
reported for whole plants (13) or plant parts treated with other
elicitors (16-21-22-23-24). The temporal relationships were
strikingly similar to those observed in plants. After 6-hour
exposure to elicitor, medicarpin concentration increased, then
reached a maximum by 48 hours, and decreased slightly over the
next 24 hours (18). A related pterocarpan (maackiain) was also
found in both whole plant and callus tissue. However, some
phytoalexins found in detached, fungus-treated clover leaflets
(Table I) were not found in elicited callus cultures (Table II).

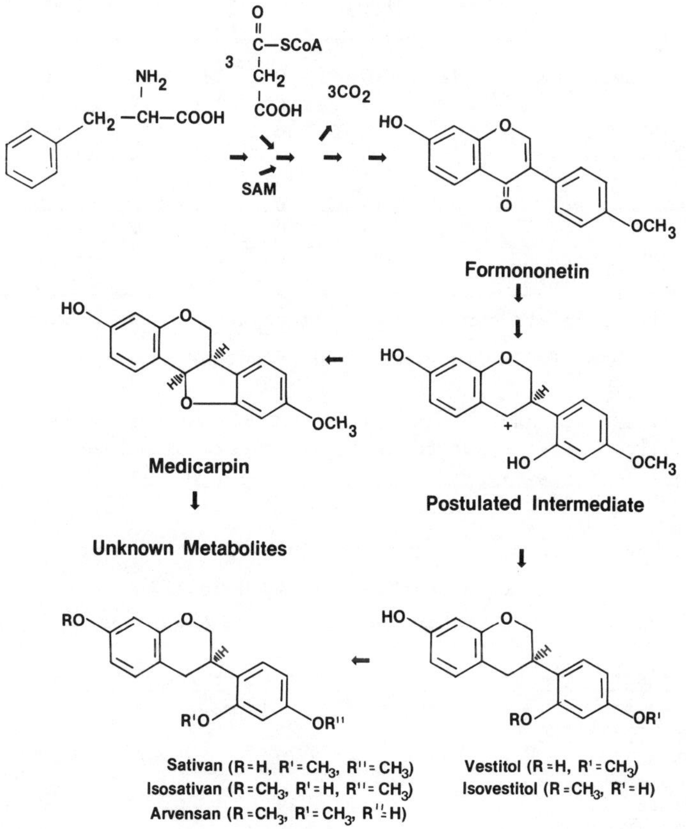

Fig. 1. Biosynthetic pathway for alfalfa and clover phytoalexins. SAM, S-adenosyl methionine.

Table I. Phytoalexins in Trifolium Detached-Leaflet Diffusates
(48 hours)[a,b]

Species	Med	Maack	4-MM	Vest	iVest	Sat	iSat
				Phytoalexin[c]			
				(μg/ml)			
Trifolium fragiferum	76	0	0	164	0	0	0
T. hybridum	70	56	2	170	0	8	10
T. incarnatum	Tr[c]	0	0	115	48	Tr	71
T. michelianum	Tr	0	0	75	0	0	0
T. medium	12	6	0	25	0	45	0
T. pratense	42	45	0	0	0	0	0
T. repens	91	0	0	0	0	0	0
T. resupinatum	Tr	0	0	109	0	0	44

[a]Treated with droplets containing Helminthosporium carbonum spores.

[b]Data from Ingham (24) with permission of author and Pergamon
Press, Ltd.

[c]Med = medicarpin, Maack = maackiain, 4-MM = 4-methoyx-maackiain,
Vest = vesitol, iVest = isovestitol, Sat = sativan, iSat =
isosativan, Tr = trace.

Table II. Phytoalexin Concentrations in HgCl$_2$-stimulated
Legume Callus Tissues (48 hours)[a]

Cultured Plant	Medicarpin[b]	Maackiain[b]
	(μg/g fresh callus)	
Trifolium fragiferum L.	99	0
T. hirtum All.	35	0
T. hybridum L.	6	0
T. incarnatum L.	17	0
T. medium L.	20	23
T. michelianum Savi	100	0
T. pratense L. cv. Kenstar	0	8
T. repens L. cv. Ladino	40	0
T. resupinatum L.	34	0
Canavalia ensiformis L.	15	0
Medicago sativa L.	30	0

[a]Data from Gustine and Moyer (20) with permission of authors and
Martinus Nijhoff.

[b]Other fungitoxic compounds were detected by thin-layer chromatography, but not in sufficient quantity for identification.

Sativan, vestitol, 4-methoxy-maackiain, isovestitol, and isosativan were not detected in callus cultures derived from legumes that normally produce them. These phytoalexins may have been present at very low levels or were not elicited by mercuric chloride.

The phytoalexin response of callus cultures derived from Ladino clover to various substances is shown in Table III. Most of the elicitors listed for the callus system have also shown elicitor activity in other plant phytoalexin-producing systems. However, most of the abiotic and biotic non-elicitors for the callus system (Table III) are elicitors in whole plant systems (see 1 and references therein).

It is clear from earlier reports (18-19-20) and the data in Tables II and III that production of medicarpin and maackiain in response to elicitors is retained in cultured cells and, most importantly, that biochemical regulation of phytoalexin accumulation is not lost.

Mechanisms Regulating Accumulation of Medicarpin in Legume Callus Cultures

Within 6 hours after exposure of jackbean callus to $\underline{P}$. chartarum spores, phenylalanine ammonia lyase (PAL) activity in crude extracts, measured spectrophotometrically, increased and reached maximum level by 24 hours (2-fold increase). During the next 24 hours, the activity decreased to about 50% of maximum (data not shown). When Sephadex $G-100$ purified fractions were assayed with ^{14}C-substrates for PAL (^{14}C-L-phenylalanine) and O-methyl transferase (OMT) (S-adenosyl methionine, ^{14}C-methyl) activities, increased levels were also found (18). The radiometric assays showed that PAL and daidzein OMT activities increased 2- and 4-fold, respectively, after 36-hour exposure to spores. OMT activities also increased 4-fold when isoliquiritigenin or genistein were used as substrates. The above enzymes and substrates are considered to be part of the medicarpin biosynthetic pathway. Caffeic acid, naringenin, and apigenin, although not in the medicarpin biosynthetic pathway, were good substrates for OMT. However, they were not methylated at a greater rate in preparations from spore-treated callus. Thus, OMT enzymes that are part of the medicarpin biosynthetic pathway seemed to be specifically stimulated.

Stimulation of PAL and other enzyme activities associated with biosynthesis of isoflavonoid phytoalexins has been demonstrated in other plants (25-26). Increased enzyme activity was accompanied by increased synthesis of mRNA's for the specific enzymes (27-28). As shown by elegant experiments conducted in Lamb's laboratory (29-30), increased de novo protein synthesis, as shown by incorporation of deuterium from D_2O into PAL protein, accounts for the increased enzyme activity during phaseollin accumulation in French bean. Recent studies utilizing genetic engineering technology demonstrated PAL and 4-coumarate CoA ligase mRNAs increased coordinately with enzyme activity in parsley suspension cultures treated with cell wall elicitor from Phytophthora megasperma (31). Similarly, chalcone synthase (CS)

Table III.   Response of <u>Trifolium</u> <u>repens</u> L. cv. Ladino Clover
Callus to Biotic and Abiotic Materials[a]

Elicitors	Non-elicitors
Biotic	**Biotic**
<u>Pithomyces</u> <u>chartarum</u> (Berk. & Curt.)	<u>Xanthomonas</u> <u>lespedezae</u>[b]
M.B. Ellis[b]	<u>Xanthomonas</u> <u>alfalfae</u>[b]
<u>Pseudomonas</u> <u>corrugata</u>[b]	<u>Botrytis</u> <u>cinerea</u>[b]
<u>Helminthosporium</u> <u>carbonum</u>	Chitin
Cystine[b]	Hexoses[d]
Oxidized glutathione[b]	Hexosamines[d]
Polygalacturonic acid fragments[c]	Amino acids[d]
Chitosan	
	Abiotic
Abiotic	Cycloheximide
Mersalyl[b]	Actinomycin D
p-chloromercuribenzoic acid (PCMBA)	Sodium cyanide
p-chloromercuribenzene sufonic acid	Sodium fluoride
(PCMBS)	Dimethyl sulfoxide
N-ethylmaleimide (NEM)	
Iodacetamide (IA)	
Diamide[b]	
Mercuric chloride (HgCl$_2$)	
Oxidized dithiothreitol[b]	
Triton X-100	

[a]All materials were tested at appropriate concentrations over a
1000-fold range in a modified Gamborg's medium (<u>19-20</u>).   Methyl
ethyl ketone (MEK) extracts were examined by silica gel TLC
(MEK-hexanes; 70:30, v/v) or by HPLC (<u>19</u>).

[b]Only tested in the callus system.

[c]Oligomer mixture from incubation of <u>Rhizopus</u> <u>stolinifer</u> endopoly-
galacteronase with polygalacturonic acid (from Dr. Charles A. West).

[d]Tested with <u>Canavalia</u> <u>ensiformis</u> (Jackbean) callus.

activity and mRNA levels were coordinately increased in soybean
seedlings inoculated with P. megasperma (32). Increased CS mRNA
was demonstrated in Phaseolus vulgaris suspension cultures in
parallel with increased activity of the enzyme after treatment
with elicitor from Colletotrichum lindemuthianum (33). Increased
mRNAs encoding for PAL, CS, and chalcone isomerase were recently
reported for the same system (34). These findings provide firm
evidence to support the proposed sequence, mRNA synthesis, protein
synthesis, and phytoalexin synthesis, in parsley and bean tissues
following stimulation with elicitors. This chain of events has
yet to be demonstrated in alfalfa or clover tissue cultures.

Cline and Albersheim (35) suggested that the primary event in
activating the biochemical processes leading to increased
concentrations of phytoalexins in plants is the binding of
elicitors to receptors in the plasma membrane. Initial evidence
for this was reported by Yoshikawa et al. (36) when they found
that soybean membrane preparations bound ^{14}C-laminaran (a β-1,3-
glucan elicitor from Phytophthora spp.). The receptor, while not
yet characterized, appears to be a protein or glycoprotein.
Small, biologically active β-1,3-glucan oligosaccharide elicitors
can also be released from larger plant or fungal polysaccharides
by β-1,3-endoglucanases found in the plant cell wall (37). Thus
it is likely that elicitors in contact with in cultured plant
cells act by mechanisms involving membrane receptors, β-1,3-
endoglucanases, or both. In addition, other membrane components
may be crucial to elicitor-stimulated responses.

Sulfhydryl Reagents and Thiols Elicit Medicarpin Accumulation in Ladino Clover Callus and Soybean Hypocotyls

Sulfhydryl (SH) reagents are excellent elicitors for stimulating
medicarpin accumulation in callus cultures (Tables II and III;
19). The SH reagents listed in Table III (abiotic elicitors) vary
in their ability to penetrate the plasma membrane and are shown in
the order of non-penetrating (mersalyl) to rapidly penetrating
($HgCl_2$) reagents. When Ladino callus was treated with mersalyl
and tested for medicarpin, the level of phytoalexin was increased
to only one-fourth of that for 6.3 mM $HgCl_2$ (Table IV). The
12-fold increase observed in the same experiment with $HgCl_2$ was
typical of the callus response to thiol reagents. All SH reagents
tested so far, with the exception of mersalyl, have induced a
12- to 15-fold increase in medicarpin concentration in callus.
These data suggest that the interior, but not the outer surface,
of the plasma membrane contains components that are sensitive to
SH reagents. The data further suggest that alteration of these
membrane components by SH reagents activates the biosynthesis of
medicarpin. Similarly, Stoessel (38) has demonstrated glyceollin
accumulation in soybean hypocotyls treated with PCMBS, PCMBA, and
NEM. He also concluded that the reactive SH groups are associated
with the plasma membrane, and they are on the outer surface rather
than in the interior membrane. Further evidence that SH groups
can regulate phytoalexin accumulation was obtained by Gustine (19)
and Stoessel (38) who found that SH groups in the plasma membrane
could be protected from SH reagents by DTT pretreatment.

Table IV. Effect of Penetrating ($HgCl_2$) and Non-penetrating
          (Mersalyl Acid) Sulfhydryl Reagents on Medicarpin
          Accumulation in Ladino Clover Callus[a]

$HgCl_2$ (mM)	Medicarpin (ug/g fr. wt.)	Mersalyl Acid (mM)	Medicarpin (ug/g fr. wt.)
0	7.6	0	2.8
2.9	93.9	1	8.9
6.3	101.2	2	12.1
12.6	82.1	4	12.5
		8	20.1
		12	19.8
		16	23.2
		20	24.9

[a]Callus tissue was mixed, divided into 12 equal portions, and
placed in incubation vials containing Gamborg's modified medium
(19-20) and indicated concentrations of abiotic elicitor. Vials
were incubated for 24 hours. Callus was separated from medium,
extracted with MEK, and medicarpin concentration determined by
HPLC analysis of extracts (19).

Gustine (19) found that prior incubation of callus for one hour in
DTT prevented stimulation of medicarpin accumulation by NEM. To
further explore this observation, callus was incubated for one
hour with 50 mM DTT and washed with DTT-free medium to remove the
reagent and any oxidized DTT. The callus was then incubated for
24 hours with 0 to 80 mM NEM. The amount of medicarpin
accumulated was determined and compared with that found in callus
elicited with NEM but not preincubated with DTT (Figure 2). These
results show that callus preincubated with DTT and washed,
required about 36 mM NEM to produce half-maximal response, whereas
callus untreated with DTT gave the same response at 10 mM NEM.
These data suggest that during pretreatment DTT converted all
accessible protein and cytoplasmic disulfides to free SH groups,
thus greatly increasing the number of reactive sites for NEM.
This means that elicitor activity of SH reagents is associated
with specific free SH groups in the cytoplasmic membrane rather
than with non-specific cell toxicity. Thus, for DTT-treated
callus, the lack of response to NEM at concentrations less than
25 mM was due to titration of SH groups, which were not associated
with the elicitor-stimulated response.
     Stoessel (38) found that 10 mM DTT reduced the amount of
elicited glyceollin in soybean hypocotyls pretreated with 10 mM
PCMBS or PCMBA. In contrast to the callus system, pretreatment
with DTT had no effect in hypocotyls exposed to 10 mM NEM. Since
the reactivity of membrane protein SH groups with thiol reagents
is variable, it is possible that some or all regulatory SH groups
are in the cytoplasm. All the SH reagents tested will react with
membrane SH groups before they enter the cytoplasm (39). Thus, IA
and $HgCl_2$ may be effective elicitors at concentrations less than

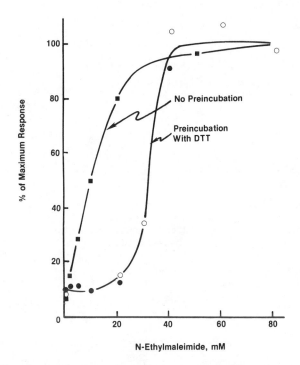

Fig. 2.   Accumulation of medicarpin in Ladino clover callus
elicited with N-ethyl maleimide.   Data are from three separate
experiments.   1) Callus was prepared as described (19) and
divided into vials containing the indicated concentrations
of NEM (one vial for each NEM concentration).   After
24-hour incubation period, callus was extracted with methyl
ethyl ketone, and medicarpin concentrations determined in
the extracts by HPLC (19); ■-■.   2) Same as experiment 1,
except the callus was incubated for 1 hour with 50 mM DTT,
and washed with DTT-free medium prior to addition of NEM;
O-O.   3) Same conditions as experiment 2; ●-●.

5 mM (19) because they react better with cytoplasmic than membraneous SH groups. The possibility still remains that cytoplasmic SH groups may be important in regulating medicarpin biosynthesis.

When callus was incubated with 50 mM diamide, a reagent that reacts preferentially with glutathione (GSH) (40) and is an elicitor (Table III), the medicarpin level after 24-hour exposure was 40% of that found for 6.3 mM $HgCl_2$ (Gustine, unpublished). I further found that oxidized GSH (GSSG) and cystine, both biological dithiols, and 2,3-dihydroxy-1,4-dithiolbutane (oxidized DTT) demonstrated elicitor activity in Ladino callus. GSSG and cystine induced medicarpin levels to about 60% of that obtained with $HgCl_2$, while oxidized DTT produced a level of about 33% of that found in $HgCL_2$-treated tissue. These results may indicate SH groups associated with the phytoalexin response are regulated by cellular dithiols.

Elicitors Stimulate Biosynthesis of Medicarpin in Ladino Clover Callus

Another aspect of regulation of medicarpin levels in Ladino callus is control of biosynthesis versus catabolism. As previously mentioned, recent evidence indicates that the primary cell response to elicitors is the transcription of new mRNAs and translation of new protein required for phytoalexin biosynthesis. This has been demonstrated in soybean elicited with P. megasperma glucan or $HgCl_2$ (41-42), in French bean elicited with heat-released elicitor from C. lindemuthianum (43), and in parsley suspension cultures elicited with P. megasperma glucan (31). Results from callus experiments are consistent with those reports. Callus was allowed to incorporate ^{14}C-L-phenylalanine or ^{14}C-acetate at different times following treatment with 6.3 mM $HgCl_2$, for one hour pulse periods. The concentrations of medicarpin and amount of ^{14}C-medicarpin were determined and are compared in Figure 3. The rates of incorporation of radioactivity into medicarpin from either precursor increased between 10 and 15 hours, and decreased thereafter. The maximum rate of incorporation occurred between 14 and 18 hours, and preceded the point at which maximum medicarpin concentration occurred by 10 to 12 hours. Because medicarpin remained at the maximum level from 20 to 50 hours, while its rate of synthesis decreased, I concluded that it was not further metabolized. Furthermore, the specific radioactivities of medicarpin synthesized from either precursor declined over time (Table V). If medicarpin was further metabolized, the specific activities would have attained a steady-state level. Thus, increased medicarpin concentration in $HgCl_2$-treated Ladino clover callus appears to be the result of increased biosynthesis, a finding observed for to other phytoalexin-accumulating systems.

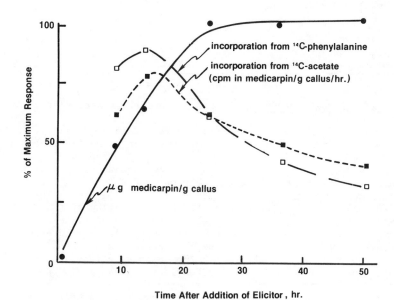

Fig. 3. Rate of incorporation of radiolabel from ^{14}C-L-phenylalanine and ^{14}C-acetate into ^{14}C-medicarpin. Sufficient callus was mixed in medium to provide about 1 g of callus per time point. At zero time, callus was added to medium containing 6.3 mM HgCl$_2$. At 8, 13, 24, 36, and 50 hours, 5 μCi of ^{14}C-L-phenylalanine (500 μCi/μmole) or 5 μCi of ^{14}C-acetate (50 μCi/μmole) was added to callus. After 1 hour, the callus was extracted and medicarpin determined (see Figure 2). Medicarpin peaks were collected during HPLC analyses and radioactivity determined by scintillation counting.

Table V. Specific Radioactivity of Medicarpin After One-hour
Incorporation of ^{14}C-L-phenylalanine or ^{14}C-acetate

Post-Elicitor Time (Hours)	Specific Radioactivity (μCi/μmole medicarpin)	
	^{14}C-L-phenylalanine	^{14}C-acetate
8-9	16.0	7.2
13-24	15.0	5.1
24-25	11.0	2.1
36-37	5.0	0.8
50-51	4.0	0.6

Biological Disulfides as Regulators of Medicarpin Accumulation

Coordinated regulation of metabolic processes is required by
plants to adapt to perceived risk of infection. The first step in
exercising control of the metabolic processes is recognition of
the perceived risk, presumably a direct consequence of an elicitor
binding to a receptor. Once recognition occurs, the processes
leading to phytoalexin accumulation would be stimulated. These
events could be activated by allosteric modification of the
receptor, alteration of the polymerization state of protein
subunits of an enzyme, or covalent attachment of an effector
molecule to the receptor or an enzyme. Besides modulation of
protein function by phosphorylation/dephosphorylation,
methylation, carboxylation, and ribosylation, changes in the
thiol/disulfide ratio may also alter enzyme activity. Gilbert
(44) recently proposed that biological disulfides may act as a
third messenger. That is, the second messenger, cAMP, induces
metabolic changes in the thiol/disulfide ratio, which acts as a
third messenger to modulate enzyme activity (44). Protein SH
groups are often important factors in enzyme activity--catalytic
function can be inhibited by SH reagents or can be maintained by
reducing reagents such as DTT or mercaptoethanol. It is clear
from this and earlier evidence (19-38), that oxidation level of SH
groups in plant cells, most likely in the plasma membrane, is
important in modulating phytoalexin biosynthesis.
     The proposal that biological dithiols act as a third
messenger to stimulate accumulation of medicarpin is highly
speculative. Current knowledge suggests several areas to
investigate to test this hypothesis. Do plant elicitor receptors
contain SH groups, and must they be oxidized or reduced for
binding to occur? Do SH reagents prevent elicitor binding to
receptors and, if so, does DTT reverse the inhibition? Another
area of interest is whether pectinolytic or cellulolytic enzymes
associated with plant membranes or cell walls are affected by SH
reagents or DTT. Yoshikawa et al. (36) found that 45 mM NEM,
27 mM iodoacetamide, and 14 mM PCMBA did not prevent binding of
^{14}C-mycolaminarin to soybean membrane fractions, while 44 mM HgCl$_2$
reduced binding to 8% of the control. They also reported that

soybean endoglucanase was inhibited more than 90% by 5 mM PCMBA or PCMBS.

Data from Gustine (19) and from the present report indicated that all of these SH reagents would have elicited medicarpin accumulation in Ladino callus, as would PCMBS, PCMBA, and NEM in soybean hypocotyls (38). It is further shown in this report that biological dithiols, oxidized GSH, and cystine were effective elicitors of medicarpin synthesis and accumulation (Table III and text). Thus, it is attractive to suggest that SH reagents act as elicitors by any one or combination of the following three mechanisms: 1) they alter the thiol/disulfide ratio and thus activate the "third messenger" pathway, 2) they react with free SH groups on a receptor protein to activate the phytoalexin response, or 3) they react with free SH groups in endoglucanases or endopolygalacturonases, stimulating the enzymatic formation of oligosacharide "endogenous" elicitors. Since preliminary data indicates SH reagents have no effect or inhibit endoglucanase from soybean (36), the last possibility appears unlikely.

The data summarized here provide preliminary evidence for regulation of phytoalexin accumulation by thiol and/or disulfide groups in the interior of the plasma membrane. A hypothesis is proposed: oxidation state of the membrane SH groups regulating the phytoalexin response is maintained by biological thiol/disulfide ratios in the cytoplasm and by accessibility of these compounds to the membrane SH moieties.

Acknowledgments

I thank Christopher S. Halliday and Leigh Jacoby for technical assistance.

Literature Cited

1. Darvill, A.G.; Albersheim, P. In "Ann. Rev. Plant Physiol." Briggs, W.R.; Jones, R.L.; Walbot, V., Eds.; Annual Reviews, Inc.: Palo Alto, 1984; Vol. 32; pp. 243-75.
2. Bailey, J.A. Physiol. Plant Pathol. 1974, 4, 477-88.
3. Rossall, S.; Mansfield, J.W.; Hutson, R.A. Physiol. Plant Pathol. 1980, 16, 135-46.
4. Sato, N.; Kitazawa, K.; Tomiyama, K. Physiol. Plant Pathol. 1971, 1, 289-95.
5. Lyon, F.M.; Wood, R.K.S. Physiol. Plant Pathol. 1975, 6, 117-24.
6. Hahlbrock, K.; Grisebach, H. In "The Flavonoids"; Harborne, J.B.; Mabry, T.J.; Mabry, H., Eds.; Academic Press: New York, 1975; Part 2; Chap. 16.
7. Martin, M.; Dewick, P.M. Phytochemistry 1980, 19, 2341-6.
8. Ingham, J.L. In "Phytoalexins"; Bailey, J.A.; Mansfield, J.W., Eds.; John Wiley and Sons: New York, 1982; Chap. 2.
9. Higgins, V.J. Physiol. Plant Pathol. 1972, 2, 289-300.
10. Higgins, V.J. Phytopathology 1978, 68, 339-45.
11. Higgins, V.J.; Smith, D.G. Phytopathology 1972, 62, 235-8.
12. Duczek, L.J.; Higgins, V.J. Can. J. Bot. 1976, 54, 2620-9.

13. Vaziri, A.; Keen, N.T.; Erwin, D.C. Phytopathology 1981, 71, 1235-8.
14. Khan, F.Z.; Milton, J.M. Physiol. Plant Pathol. 1975, 7, 179-87.
15. Flood, J.; Milton, J.M. Physiol. Plant Pathol. 1982, 21, 97-104.
16. Kahn, F.Z.; Milton, J.M. Physiol. Plant Pathol. 1978, 13, 215-21.
17. Mansfield, J.W. In "Phytoalexins"; Bailey, J.A.; Mansfield, J.W., Eds.; John Wiley and Sons: New York, 1982; Chap. 8.
18. Gustine, D.L.; Sherwood, R.T.; Vance, C.P. Plant Physiol. 1978, 61, 226-30.
19. Gustine, D.L. Plant Physiol. 1981, 68, 1323-6.
20. Gustine, D.L.; Moyer, B.G. Plant Cell Tiss. Org. Cult. 1982, 1, 255-63.
21. Kahn, F.Z.; Milton, J.M. Physiol. Plant Pathol. 1979, 14, 11-7.
22. Cruickshank, I.A.M.; Veeraraghavan, J.; Perrin, D.R. Aust. J. Plant Physiol. 1974, 1, 149-56.
23. Cruickshank, I.A.M.; Spencer, K.; Mandryk, M. Physiol. Plant Pathol. 1979, 14, 71-6.
24. Ingham, J.L. Biochem. Syst. Ecol. 1978, 6, 217-23.
25. Lamb, C.J.; Lawton, M.A.; Taylor, S.J.; Dixon, R.A. Ann. Phytopathol. 1980, 12, 423-33.
26. Dixon, R.A.; Prakash, M.D.; Lawton, M.A.; Lamb, C.J. Plant Physiol. 1983, 71, 251-6.
27. Hadwiger, L.A.; Schwochau, M.E. Plant Physiol. 1971, 47, 346-51.
28. Yoshikawa, M.; Yamauchi, K.; Masago, H. Plant Physiol. 1978, 61, 314-7.
29. Lamb, C.J.; Dixon, R.A. F.E.B.S. Lett. 1978, 94, 277-80.
30. Dixon, R.A.; Lamb, C.J. Biochim. Biophys. Acta 1979, 586,453-63.
31. Kuhn, D.H.; Chappell, J.; Boudet, A.; Hahlbrock, K. Proc. Nat. Acad. Sci. U.S.A. 1984, 31, 1102-6.
32. Schmelzer, E.; Borner, H.; Grisebach, H.; Ebel, J.; Hahlbrock, K. FEBS Lett. 1984, 172, 59-63.
33. Ryder, T.B.; Cramer, C.L.; Bell, J.N.; Robbins, M.P.; Dixon, R.A.; Lamb, C.J. Proc. Nat. Acad. Sci. U.S.A. 1984, 81, 5724-8.
34. Cramer, C.L.; Ryder, T.B.; Bell, J.N.; Lamb, C.J. Science 1985, 227, 1240-3.
35. Cline, K.; Albersheim, P. Plant Physiol. 1981, 68, 221-8.
36. Yoshikawa, M.; Keen, N.T.; Wang, M-C. Plant Physiol. 1983, 73, 497-506.
37. Keen, N.T.; Yoshakawa, M. Plant Physiol. 1983, 71, 460-5.
38. Stoessel, P. Planta 1984, 160, 314-9.
39. Rothstein, A. In "Curr. Topics Membr. Struct." Bronner, F.; Kleinzeller, A., Eds.; Academic Press: New York, 1970; Vol. 1; pp. 135-176.
40. Kosower, E.M.; Correa, W.; Kinon, B.J.; Kosower, N.S. Biochim. Biophys. Acta. 1972, 264, 39-44.
41. Moesta, P.; Grisebach, H. Arch. Biochem. Biophys. 1981, 211, 39-43.

42.   Moesta, P.; Grisebach, H.   Arch. Biochem. Biophys. 1981, 212, 462-7.
43.   Lawton, M.A.; Dixon, R.A.; Hahlbrock, K.; Lamb, C.   Eur. J. Biochem. 1983, 129, 593-601.
44.   Gilbert, H.F.   J. Biol. Chem. 1982, 257, 12089-91.

RECEIVED August 9, 1985

ROLES OF ALLELOCHEMICALS IN PLANT-INSECT INTERACTIONS

6

Polyhydroxy Plant Alkaloids as Glycosidase Inhibitors and Their Possible Ecological Role

L. E. Fellows, S. V. Evans, R. J. Nash, and E. A. Bell

Jodrell Laboratory, Royal Botanic Gardens, Kew, TW9 3DS, United Kingdom

Polyhydroxyalkaloids which structurally resemble monosaccharides have been found in a variety of organisms,
including higher plants. Many are potent inhibitors of
glycosidase activity in insects, mammals and microorganisms and it is suggested that these properties
contribute to the natural chemical defences of those
plants in which they accumulate.

In recent years, representatives of 3 new classes of alkaloid have
been isolated from both plants and micro-organisms, namely polyhydroxy derivatives of indolizidine, pyrrolidine and piperidine.
Their relatively late discovery is probably due to their failure to
react with most colorimetric reagents used to detect alkaloids, such
as Dragendorff reagent, but most do react with ninhydrin to give a
yellow or brown colour.
    Evidence is growing that these compounds are widely distributed
in plants, representatives having now been isolated from four separate families, viz. Moraceae (1), Leguminosae (2-8), Polygonaceae (9)
and Aspidiaceae (10). In some legumes we have observed that these
alkaloids may accumulate as 2% or more of the plant's dry weight (11)
which suggests that they have an ecological role. Many other types
of alkaloid are of course known to be effective deterrents to preda-
tion, particularly by mammals. The unhydroxylated piperidine alkaloids of Hemlock (Conium maculatum) poison the mammalian nervous
system, as Socrates well knew, but when a polyhydroxy piperidine
(Figure 1) was isolated at Kew from the legume Lonchocarpus sericeus
(2), no activity of this nature was detected. 1 is an isomer of
2-hydroxymethyl-3,4,5-trihydroxypiperidine, which can also be
described as 1,5-dideoxy-1,5-imino-D-mannitol; it is in fact an analogue of 1-deoxymannose in which the ring oxygen is replaced by
nitrogen. This suggested that 1 might mimic the action of mannose
and thereby inhibit carbohydrase, particularly mannosidase, activity.
    We were unaware at this time that Japanese workers had isolated
from certain mulberry (Morus) spp. a compound (Figure 2) being an
epimer of 1 at C5, which they named moranoline (1). It was later
also found in bacteria (12-13) and described as deoxynojirimycin, the
name which now prevails. This compound, 2, is an analogue of

0097–6156/86/0296–0072$06.00/0

1-deoxyglucose, being an epimer of $\underline{1}$ at C2 if each molecule is considered as a monosaccharide, and inhibits both α - and β -glucosidases
from varied sources, including fungal trehalase ($\underline{12\text{-}15}$). It also
inhibits the postprandial rise in blood glucose if administered to
rats with food and a patent was filed on $\underline{2}$ as an antidiabetic agent
($\underline{16}$). Recently it has been shown to be a potent inhibitor of mouse
gut digestive disaccharide activity, particularly with substrates
incorporating an α -linked glucopyranosyl residue ($\underline{17}$). The mannose
analogue $\underline{1}$ inhibits α -mannosidase from various sources, and also
bovine epididymis α -L-fucosidase ($\underline{14}$, $\underline{18\text{-}19}$). No glucosidase inhibition has been reported.

An alkaloid closely related to both $\underline{1}$ and $\underline{2}$, but lacking a
hydroxy group at C5 (C2 if considered as a sugar) was detected in
buckwheat (<u>Fagopyrum esculentum</u>) and named fagomine (Figure $\underline{3}$), ($\underline{9}$).
Recently we have isolated a glucoside of fagomine, 4-O-(β-D-gluco-
pyranosyl)-fagomine, ($\underline{5}$) from the legume <u>Xanthocercis zambesiaca</u>.
Neither fagomine nor its glucoside showed any inhibition of a range
of glucosidases tested which confirms previous findings of the importance of the hydroxy group at C2 (piperidine C5) in catalysis ($\underline{20}$).

It was realised that a polyhydroxy pyrrolidine, DMDP (Figure $\underline{4}$),
isolated from <u>Derris elliptica</u> ($\underline{21}$), was an analogue of β -D-fructo-
furanose, and at Kew we were able to isolate $\underline{4}$ from related legumes
(<u>Lonchocarpus</u> spp.) in sufficient quantity to carry out a range of
biological tests. (It has also been synthesised and its absolute
configuration confirmed as 2R-,5R-dihydroxymethyl-3R,4R-dihydroxy-
pyrrolidine) ($\underline{22}$). DMDP also proved a potent glucosidase inhibitor,
but its action differs from that of $\underline{2}$ in several respects: for
example, while $\underline{4}$ inhibits yeast invertase (50% inhibition at 5.25 x
10^{-5}M), $\underline{2}$ has no effect at 5 x 10^{-3}M. ($\underline{14}$). Conversely, $\underline{2}$ is a more
effective inhibitor than $\underline{4}$ of mouse gut invertase (50% inhibition at
7.6 10^{-8}M and 4.6 x 10^{-5}M respectively) ($\underline{17}$). Inhibition of lysosomal β -D-mannosidase and of a mannosidase of intestinal epithelial
glycoprotein processing by $\underline{4}$ has also been reported ($\underline{23\text{-}24}$). DMDP
also inhibits insect trehalase ($\underline{14}$). Since trehalose, 1-(α -D-gluco-
pyranosyl-α-D-glucopyranoside) is an important storage carbohydrate
in insects but not in mammals, inhibition of insect trehalase could
theoretically provide the basis of a relatively safe insecticide.

At Kew we have been particularly interested in the chemical
basis of the interaction between bruchid beetles and their legume
hosts. Bruchid beetles normally lay eggs on green pods; the larvae
penetrate pod and testa, develop in the seed and emerge as adults.
Crop grain legumes have few natural chemical defences of their own
and are potential prey for many bruchid spp. but, provided the
bruchid retains its requirement for a green pod for oviposition, a
low level of contamination at harvest is no problem since emerging
adults do not multiply. This requirement has been lost, however, by
certain spp. which can breed in storage and are now major economic
pests ($\underline{25}$). Wild legumes are potential sources of anti-bruchid
chemicals which might be used for crop protection. When $\underline{4}$ was
incorporated into artificial 'seeds' made of cow pea flour packed in
gelatine capsules, levels of only 0.03% were lethal to larvae of the
pest bruchid <u>Callosobrusus maculatus</u> ($\underline{26}$). (Fagomine and its glucoside had no effect, even at 1%). In contrast, mice fed a diet containing 5% $\underline{4}$ for one week showed no ill effects other than a slight

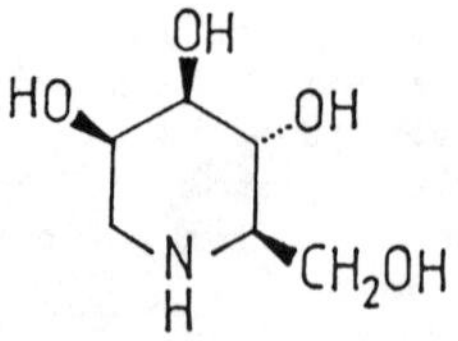

Figure 1.   Deoxymannojirimycin

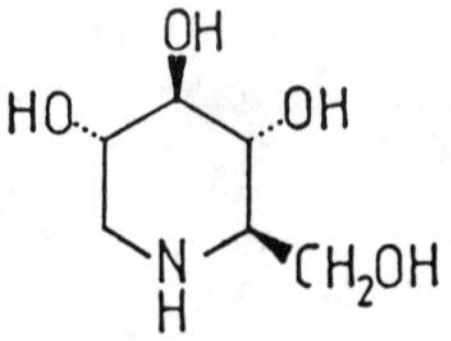

Figure 2.   Deoxynojirimycin

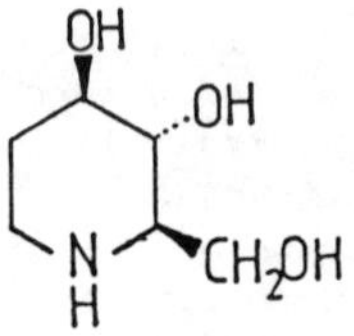

Figure 3.   Fagomine

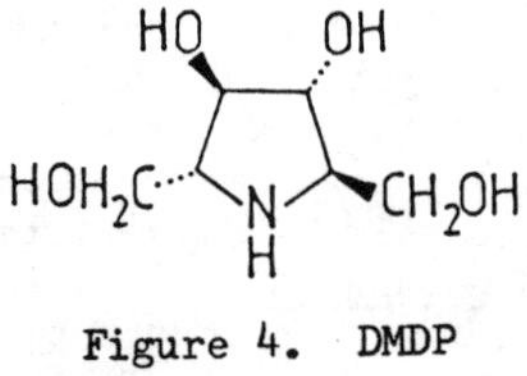

Figure 4.   DMDP

weight loss and recovered completely when returned to a normal diet
(27). α-Glucosidase activity in $\underline{C}$. maculatus larval gut was
strongly inhibited by $\underline{4}$ (K_I 3.6 x 10^{-7}) (28) but that of mouse gut,
(e.g. K_I for maltase c. 10^{-4}M) far less so (17).

The biochemical mechanisms by which insects select food plants
are not fully understood. It has been suggested that an insect
glucosidase may be involved which, when in contact with the plant,
releases free sugars which trigger the feeding response (28). A
glucosidase inhibitor might therefore be expected to be an effective
feeding deterrent. Blaney et al. (29) presented 5th instar nymphs
of the acridids Schistocerca gregaria (polyphagous) and Locusta
migratoria (oligophagous) with glass fibre discs impregnated with
known phagostimulants for locusts. On discs containing 5% sucrose,
for example, DMDP levels as low as 0.001% were sufficient to reduce
feeding, yet when locusts were force-fed (1mg/g body weight) there
was no discernible effect. Higher concentrations of DMDP were
required to deter feeding in final instar larvae of two Spodoptera
spp. (armyworms), the oligophagous $\underline{S}$. exempta being more easily
deterred than the polyphagous $\underline{S}$. littoralis. Both species had high
mortality when presented with sucrose-laden discs containing 1% DMDP;
discs with 0.01% DMDP caused mortality only in $\underline{S}$. littoralis owing to
its having consumed more DMDP-laden food. Results of electrophysiological studies on the Spodoptera gustatory sensilla could be interpreted to indicate that DMDP was competing with a glucosidase receptor site, particularly in $\underline{S}$. exempta, but the neurone firing in a
dose-dependent manner responded similarly to alkaloids. The authors,
in conclusion, favoured the view that the deterrency of DMDP derived
more from its properties as an alkaloid than as a sugar analogue.
(Similar studies with $\underline{2}$ suggest that it is far less effective as a
feeding deterrent than $\underline{4}$) (30).

A compound related to $\underline{4}$, being an isomer of 2-hydroxymethyl-3,4-
dihydroxypyrrolidine having the D-arabinose configuration, (alternatively 1,4-dideoxy-1,4-imino-D-arabinitol), (Figure $\underline{5}$), has been
isolated from the legume Angylocalyx boutiqueanus and found to be a
more potent inhibitor of yeast α-glucosidase than either $\underline{2}$ or $\underline{4}$ (6,8).
(An isomer purported to have the D-xylose configuration has been
found in the fern Arachnoides standishii but other authors claim that
it is identical with $\underline{5}$) (10, 31). The related (2R, 3S)-2-hydroxy-
methyl-3-hydroxypyrrolidine (Figure $\underline{6}$), has been isolated from the
legume Castanospermum australe (7). It is a very weak inhibitor of
some glucosidases, but at 0.1% did deter Schistocerca gregaria from
feeding on sucrose-laden discs.

Two polyhydroxy indolizidine alkaloids, swainsonine (Figure $\underline{7}$)
and castanospermine (Figure $\underline{8}$), are known to occur in legumes ($\underline{3}$, $\underline{32}$,
$\underline{4}$). Swainsonine is responsible for the toxic effects observed in
animals grazing swainsonine-accumulating plants (33). An inhibitor
of α-mannosidase, it induces a lysosomal storage disease in animals
which resembles a genetic disorder characterised by a lack of α-
mannosidase, and is being used to induce these conditions in normal
cell lines (34). It has also been found in a fungus (35). Castanospermine, $\underline{8}$, from Castanospermum australe, inhibits glucosidases
from several sources (36-37) but differs from $\underline{2}$ and $\underline{4}$ in specificity:
it has no effect, for example, on yeast α-glucosidase but does
inhibit emulsin β-glucosidase. We have found it to be lethal to

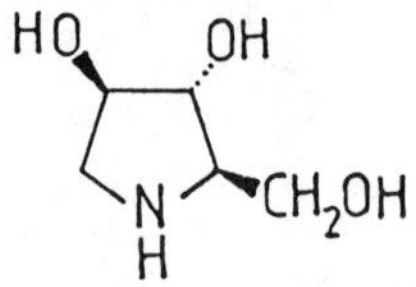

Figure 5.   D-AB1

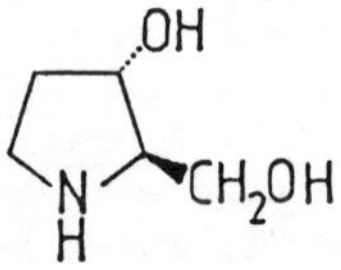

Figure 6.   CYB-3

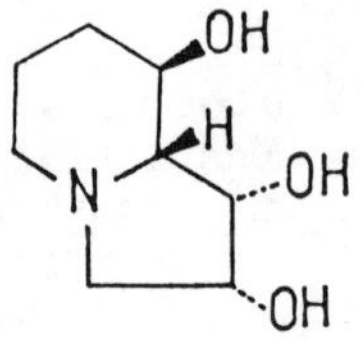

Figure 7.   Swainsonine

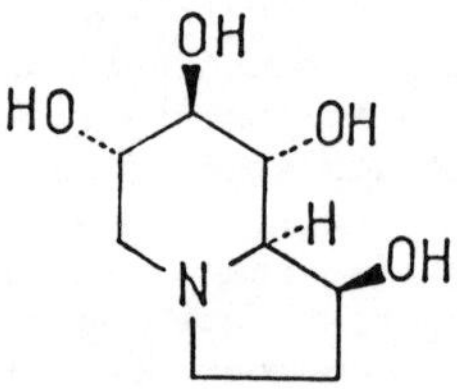

Figure 8.   Castanospermine

larvae of <u>Callosobrusus</u> <u>maculatus</u> and the flour beetle <u>Tribolium</u>
<u>confusum</u> at 0.03% and 0.1% of the diet respectively. In preliminary
studies, <u>Locusta</u> <u>migratoria</u> was inhibited from feeding on sucrose-
laden discs incorporating 0.02% alkaloid, whereas <u>Schistocerca</u>
<u>gregaria</u> was unaffected by 1%.

In summary, there is now considerable evidence that polyhydroxyalkaloids in plants can exert deterrent or deleterious effects on
potential predators. Many aspects of this interaction merit further
study, such as effects on micro-organisms and other plant spp. In
pilot studies in this laboratory, we have observed inhibition of
vegetative growth and spore germination in certain fungi, pollen
germination and IAA stimulated elongation of wheat coleoptile segments by certain of these alkaloids. On germination of <u>Lonchocarpus</u>
<u>sericeus</u> and <u>Castanospermum</u> <u>australe</u>, there is selective excretion of
hydroxyalkaloids for several days. Extensive comparative studies
have been hindered by limited supplies of purified compounds but,
with synthetic material increasingly available (<u>8</u>, <u>19</u>, <u>22</u>, <u>38-40</u>),
these difficulties should be reduced.

Conclusions

1.   Polyhydroxyalkaloids, many of which are potent glycosidase
inhibitors, are probably widespread in nature and may contribute to
the chemical defence of those plants in which they accumulate.
2.   Those polyhydroxyalkaloids which are less inhibitory to mammalian glycosidases than those from other organisms may find application
in crop protection formulations.
3.   Polyhydroxyalkaloids have potential as research tools for
probing the chemistry and biology of glycosidase reactions.

Acknowledgments

We thank Prof. D. H. Janzen for plant material, Ms C. Doherty, Mrs
P. Churcher and P. W. Smith for technical help, and the S.E.R.C. and
N.E.R.C. for awards (to S.V.E. and R.J.N.).

Literature Cited

1.   Yagi, M.; Kouno, T.; Aoyagi, Y.; Murai, H. <u>Nippon</u> <u>Nogei</u> <u>Kagaku</u>
     <u>Kaishi</u> 1976, 50, 571.
2.   Fellows, L. E.; Bell, E. A.; Lynn, D. G.; Pilkiewicz, F.; Miura,
     I.; Nakanishi, K. <u>J.C.S.</u> <u>Chem.</u> <u>Comm.</u> 1979, 977.
3.   Dorling, P. R.; Huxtable, C. R.; Colegate, S. <u>Biochem.</u> <u>J.</u> 1980,
     191, 649.
4.   Hohenschutz, L. D.; Bell, E. A.; Jewess, P. J.; Leworthy, D. P.;
     Pryce, R. J.; Arnold, E.; Clardy, J. <u>Phytochem.</u> 1981, 20, 811.
5.   Evans, S. V.; Hayman, A. R.; Fellows, L. E.; Shing, T. K. M.;
     Derome, A. E.; Fleet, G. W. J. <u>Tetrahedron Lett.</u> 1985, 26, 1465.
6.   Nash, R. J.; Bell, E. A.; Williams, J. M. <u>Phytochem.</u> 1985,
     24, 1620.
7.   Nash, R. J.; Bell, E. A.; Fleet, G. W. J.; Jones, R. H.;
     Williams, J. M. <u>J.C.S.</u> <u>Chem.</u> <u>Comm.</u> 1985, in press.
8.   Fleet, G. W. J.; Nicholas, S. J.; Smith, P. W.; Evans, S. V.;
     Fellows, L. E.; Nash, R. J. <u>Tetrahedron Lett.</u> 1985, 26, 3127.

9. Koyama, M.; Sakamura, S. Agr. Biol. Chem. 1974, 38, 1111.
10. Furukawa, J.; Okuda, S.; Saito, K.; Hatanaka, S. I. Phytochem. 1985, 24, 593.
11. Evans, S. V.; Fellows, L. E.; Bell, E. A. Biochem. Syst. Ecol. 1985, in press.
12. Schmidt, D. D.; Frommer, W.; Müller, L.; Truscheit, E. Naturwissen 1979, 66, 584.
13. Murao, S.; Miyata, S. Agr. Biol. Chem. 1980, 44, 219.
14. Evans, S. V.; Fellows, L. E.; Shing, T. K. M.; Fleet, G. W. J. Phytochem. 1985, in press.
15. Elbein, A. D. C.R.C. Critical Reviews in Biochem. 1984, 16, 21.
16. Murai, H.; Ohata, K.; Enomoto, Y.; Yoshikuni, Y.; Kono, T.; Yagi, M. German pat. 2656602, 1977; Chem. Abstr. 1977, 87, 141271.
17. Scofield, A. M., personal communication.
18. Fuhrmann, U.; Bause, E.; Legler, G.; Ploegh, H. Nature 1984, 307, 755.
19. Legler, G.; Juelich, E. Carbohyd. Res. 1984, 128, 61.
20. Brockhaus, M.; Dettinger, H. M.; Kurz, G.; Lehmann, J.; Wallenfels, K. Carbohyd. Res. 1978, 69, 264.
21. Welter, A.; Dardenne, G.; Marlier, M.; Casimir, J. Phytochem. 1976, 25, 747.
22. Fleet, G. W. J.; Smith, P. W. Tetrahedron Lett. 1985, 26, 1469.
23. Cenci di Bello, I.; Dorling, P.; Evans, S. V.; Fellows, L. E.; Winchester, B. Biochem. Soc. Trans. 1985, in press.
24. Romero, P. A.; Friedlander, P.; Fellows, L. E.; Evans, S. V.; Herscovics, A. FEBS Lett. 1985, in press.
25. Birch, N.; Southgate, B. J.; Fellows, L. E. In "New Plants for Arid Lands"; Wickens, G. E.; Field, D. V.; Goodin, J. R., Eds.: George Allen & Unwin/Royal Botanic Gardens, Kew, 1985; p. 303.
26. Evans, S. V.; Gatehouse, A. M. R.; Fellows, L. E. Ent. exp. appl. 1985, in press.
27. Janzen, D. H., personal communication.
28. Hansen, K. In: 25 Mosbacher Colloquium Ges. Biol. Chem.; Jaenicke, L., Ed.; Springer Verlag, 1974; p. 207.
29. Blaney, W. M.; Simmonds, M. S. J.; Evans, S. V.; Fellows, L. E. Ent. exp. appl. 1984, 36, 209.
30. Simmonds, M. S. J., personal communication.
31. Jones, D. W. C.; Nash, R. J.; Bell, E. A.; Williams, J. M. Tetrahedron Lett. 1985, in press.
32. Molyneux, R. J.; James, L. F. Science 1982, 216, 190.
33. Tulsiani, D. R. P.; Broquist, H. P.; James, L. F.; Touster, O. Arch. Biochem. Biophys. 1984, 232, 76.
34. Cenci di Bello, I.; Dorling, P.; Winchester, B. Biochem. J. 1983, 215, 693.
35. Schneider, M. J.; Ungemach, F. S.; Broquist, H. P.; Harris, T. M. J. Am. Chem. Soc. 1982, 104, 6863.
36. Saul, R.; Ghidoni, J. J.; Molyneux, R. J.; Elbein, A. D. Proc. Nat. Acad. Sci. 1985, 82, 93.
37. Saul, R.; Molyneux, R. J.; Elbein, A. D. Arch. Biochem. Biophys. 1984, 230, 668.
38. Fleet, G. W. J.; Gough, M. J.; Smith, P. W. Tetrahedron Lett. 1984, 25, 1853.
39. Bernotas, R. C.; Ganem, B. Tetrahedron Lett. 1984, 25, 165.
40. Kinast, G.; Schedel, M. Angew. Chem. 1981, 20, 805.

RECEIVED October 3, 1985

Cotton Terpenoid Inhibition of *Heliothis virescens* Development

Robert D. Stipanovic[1], Howard J. Williams[2], and Laurel A. Smith[3]

[1]National Cotton Pathology Research Laboratory, USDA, Agricultural Research Service, College Station, TX 77841
[2]Department of Entomology, Texas A&M University, College Station, TX 77843
[3]Department of Statistics, Texas A&M University, College Station, TX 77843

Some genera of the Malvaceae family, including Cienfuegosia, Gossypium (cotton), Kokia, and Thespesia, contain subepidermal pigment glands from which the phenolic sesquiterpenoid aldehyde dimer, gossypol, has been isolated (1). Elegant work by Roger Adams and co-workers resulted in the elucidation of the structure of gossypol (**G**) as shown in Figure 1 (2). Due to its toxicity to monogastric animals, **G** has been extensively studied (3,4). Reports that **G** acts as a male antifertility agent (5-7) have resulted in a new series of investigations in this area (8). **G** reportedly affects the activities of some membrane-bound mitochondrial enzymes (9), uncouples mitochondrial oxidative phosphorylation (10), inhibits (Na^+-K^+)-ATPase (11), inhibits anion transport in red blood cells (12), and alters bilayer conductance and proton permeability in model membranes (13).

Distribution of Terpenoid Aldehydes in Tissues

Seed. In cotton, pigment glands are found throughout "glanded" seed. Processing this seed produces a meal containing **G** ranging in concentrations from less than 0.02 to 0.22% depending on seed type and the method used for extracting the oil (14).

Roots and Stems. In addition to **G**, a group of related terpenoids are found in glands and in epidermal and scattered cortical cells of young roots (15). In older roots the glands occur throughout the phelloderm of the root bark. The terpenoids in epidermal cells may exude from the root surface (16). Several months after planting, **G** also is found in individual xylem ray cells in the wood (17). In diseased G. arboreum, deoxydesoxyhemigossypol (**dDHG**), desoxyhemigossypol (**dHG**), and hemigossypol (**HG**) are found in infected cambial tissue (Figure 1) (18). In diseased G. barbadense (Pima cotton), the methylated compounds desoxyhemigossypol 6-methyl ether (**dMHG**), hemigossypol-6-methyl ether (**MHG**) (19), gossypol-6-methyl ether (**MG**), and gossypol-6,6'-dimethyl ether (**DMG**) (20) are produced in addition to the compounds produced by G. arboreum. In G. hirsutum (commercial Upland cotton) the methylated derivatives also

0097-6156/86/0296-0079$06.00/0

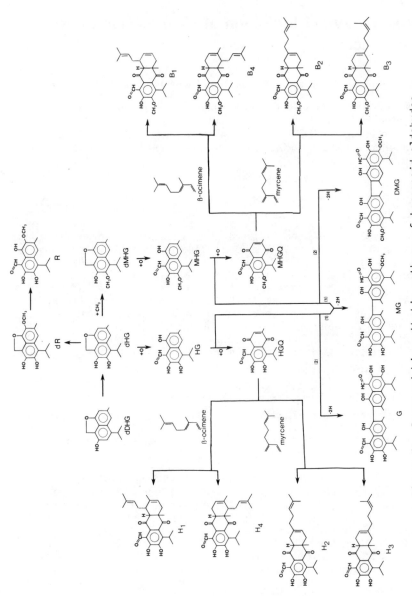

Figure 1. Proposed biosynthetic pathway of terpenoid aldehydes in cotton.

are produced, but in small amounts, in response to infection by plant pathogens. This induced synthesis has been reviewed (17). A peroxidase enzyme is responsible for the dimerization of HG to G (21), and presumably for the synthesis of MG and DMG, respectively, from the dimerization of HG and MHG and MHG only (Fig. 1).

Foliar Tissue. Pigment glands are also found in foliar parts of the cotton plant, but the terpenoid biosynthesis follows a different pathway. In young flower buds the principal terpenoid aldehyde is G (25); however, in young bolls (26) and leaves (23), G is a minor component, since HG is apparently rapidly oxidized to hemigossypolone (HGQ) (27) rather than dimerized by a peroxidase enzyme. HGQ is a good dienophile and undergoes a Diels-Alder reaction with 1,4-dienes present in the glandular oil. In G. hirsutum heliocides H_1 (H_1) and H_4 (H_4) (28), and heliocides H_2 (H_2) (29) and H_3 (H_3) (30) are formed respectively from the reaction of HGQ with β-ocimene and myrcene (Fig. 1). In G. barbadense only β-ocimene is present in the glandular oil and thus only H_1 and H_4 are formed from HGQ and heliocides B_1 (B_1) and B_4 (B_4) are formed from MHGQ (Fig. 1) (31). All eight heliocides are found in progeny from crosses between G. barbadense and G. hirsutum (24, 32). In G. barbadense both HG and MHG are synthesized; oxidation leads to HGQ and hemigossypolone-6-methyl ether (MHGQ) (31), respectively. A Schiff base, gossyrubilone, formed between HGQ and isopentyl amine has also been isolated from G. hirsutum (24). In G. raimondii, the terpenoid aldehyde synthesis is modified, and a methoxyl group is introduced at the 2-position to give raimondal (R) presumably through an intermediate such as desoxyraimondal (dR) (Fig. 1) (33).

Biological Activity of Terpenoid Aldehydes

Antibiosis. In 1959, McMichael reported the development of a cotton line which contains no pigment glands in the foliar plant parts (34), and entomologists found these glandless lines to be more susceptible than glanded ones to phytophagous insects such as Heliothis spp. In general, larval development is faster on glandless cotton than on comparable glanded cottons (24, 35, 36, 37). H. virescens larvae reared on artificial diets containing G grew significantly slower than on G free diets (36). These feeding experiments have been duplicated at several laboratories with many of the compounds in Fig. 1. The results from these laboratories are given in Table 1 as the percentage of compound required to reduce larval weight by 50% (ED_{50}).

Hormetic Effect. In order to establish a base line growth and response pattern for our experiments at College Station, TX., H. virescens larvae were fed G in artificial diet at several dosages. In these preliminary experiments, we found large variations in larval growth, especially at lower gossypol dosages. Larvae feeding on diets containing low levels of G were actually larger than larvae feeding on a gossypol free control diet.

Table I. Inhibition of H. virescens larval growth by cotton
          terpenoid aldehydes, ED_{50} as percent of diet.

Constituent	ED_{50} as % Constituent in Diet		
	Miss. State, MS[a]	Albany, CA[b]	Brownsville, TX[c]
G	0.11	0.12	0.04
H_1	-	0.12	0.10
H_2	-	0.13	0.46
H_3	-	-	0.16
HGQ	0.08	0.08	0.29
B_1	-	-	0.20[e]
B_2 & B_3 [d]	-	-	N.E.
MHGQ	-	-	N.E.

[a]ref. 27; [b]ref. 38; [c]ref. 22; [d]added to the diet as a mixture of B_2
(67%) and B_3 (33%); [e]no effect.

Based on these preliminary studies, we examined the growth and
development of H. virescens larvae fed an artificial diet containing
G under the following protocol:
1. Feed G at enough different dosages to construct an accurate
    growth-response curve.
2. Weigh larvae individually and use a sufficient number of
    larvae to observe statistically significant variations in
    growth.
3. Examine the suitability of probit analysis for modeling the
    growth-response curve, and derive an alternative mathematical
    model for such analyses if necessary.
    Thus, G (as the gossypol acetate complex) was fed to neonate H.
virescens larvae at concentrations ranging from 0-250 mg/200 mL diet
in 25 mg increments (0-1.25 mg/mL in 0.125 mg increments). One
hundred larvae were grown at each dosage for seven days. The
experiment was repeated.
    The results from one experiment are shown in Fig. 2. Maximal
larval weight was observed at a dosage of 25 mg G/200 mL diet. The
second experiment gave similar results with a maximum at 50 mg/200 mL
diet. Such a growth curve is described as a hormetic effect (39).
Hormesis is the name given to the stimulatory effect caused by low
levels of potentially toxic agents. As Figure 2 shows, growth was
reduced at 37.5 mg G/200 mL diet and above as compared to the
control. At the higher concentrations growth virtually stopped.
    Figure 2 also shows that variations in weights were
substantially less at the higher dosages. This change in the size of
the standard deviations observed in Figure 2 is apparently due to the
presence of larvae at several different instars within the low
dosage groups. Figure 3 is a log-linear graph of the weight
distribution of 100 larvae from each of the five instars reared on
the control diet, the specific instar being determined by measurement
of the larval head capsule. The mean weights are linear and thus
represent a logarithmic growth rate in agreement with other work
(40). The data in Figure 3 may be used to estimate the percentage of

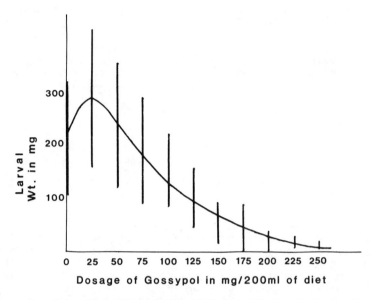

Figure 2.   Mean (± 1 SD) weights of 7 day old H. virescens larvae
fed given amounts of gossypol in Vanderzant wheat germ diet.

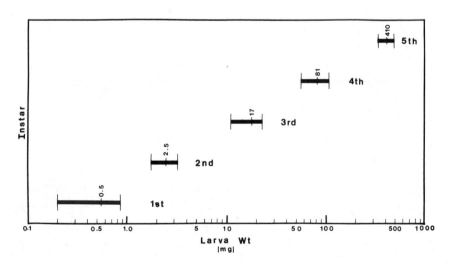

Figure 3.   Log-linear graph of larval weight for five instars of
H. virescens.

larvae at different instars. The calculated results for five **G** dosages are given in Table II. The 1st, 2nd, and 3rd instars are combined into one group.

Table II. Development of H. virescens larvae fed on artificial diets containing varying dosages of gossypol (**G**) for seven days.

Gossypol conc. (mg/mL)	1st, 2nd, 3rd Instar	% of Larvae 4th Instar	5th Instar
0	3	31	66
0.125	5	19	76
0.500	3	68	29
1.00	83	17	0
1.25	99	1	0

These results indicate that **G** retards larval development and thus extends the length of time required to reach pupation. This lengthening of development time gives high glanded cotton certain advantages over glandless or low glanded plants. Larvae feeding on high glanded cotton plants may be exposed to adverse environmental effects, disease, parasites and predators for a longer period of time, reducing their chance of survival to adulthood and also limiting the number of generations which are produced during a growing season.

The presence or absence of glands also dramatically affects the feeding selection site of H. virescens on cotton. In glanded cottons, larvae feed predominately inside flower buds and bolls where the concentration of glands is lower as compared to young leaves. In glandless plants, the leaves become the preferred feeding site rather than fruit (41). This is an important consideration when evaluating Heliothis damaged plants in the field.

Inadequacies of Probit Analysis: A Proposed New Model. Probit analysis is a recognized method of evaluating insecticidal activity by determining the dose which causes death of a given proportion of a population. The value most commonly quoted is the LD_{50} or the dose required to kill 50% of the insect population under specified conditions (42). The probit model is given by the equation:

$$w = \alpha_0 \{1 - \Phi[(Dose - \alpha_1)/\alpha_2]\}$$

where w is the mean larval weight at a given dosage, α_0 is the mean larval weight of the control, α_1 is the LD_{50} value, Φ is the cumulative normal probability distribution function and α_2 is a scaling factor. This analysis has been adapted to allelochemics that may not in themselves be insecticidal. The concentration of allelochemic needed to retard growth by 50% as compared to a control is referred to as the ED_{50} value. In general, if the probit model adequately fits the observed results, the mean larval weight will follow a sigmoidal S-shaped curve, decreasing monotonically as dosage increases. The ED_{50} value will be the dosage at which the mean larval weight is half of the mean larval weight for the controls.

The probit model does not adequately describe the growth-response curve produced when **G** is fed to **H.** virescens larvae due to the hormetic effect observed at lower dosages, which causes a mean weight increase instead of the steady decrease required by probit model. We propose a new exponential/probit model described by the equation:

$$w = \alpha_0 \{1 - \Phi[(Dose - \alpha_1)/\alpha_2]\} + \alpha_3 \cdot e^{\alpha_4 \cdot Dose}$$

where w, α_0, α_1, α_2 and Φ have the same meaning as in a probit analysis and α_3 and α_4 are scaling factors (43). This equation retains the probit portion of the curve and adds a positive exponential term which increases at low dosages, but is nearly constant at high dosages. The resulting dose-response function shows an increase in mean larval weight at very low dosages, followed by a probit-like decrease in weight at higher dosages. The hormetic/probit model has the advantage of reproducing the hormetic effect, and it gives an ED_{50} value which is directly related to other results determined by simple probit analysis. A comparison of the experimental and calculated values for one of the **G** feeding experiments is given in Figure 4. The second experiment gave similar results. ED_{50} values for the two experiments were 117mg/200mL of diet and 125 mg/200mL of diet or 0.06% of the diet, which is in good agreement with earlier results from Brownsville, TX (Table I). The hormetic maxima in the two experiments were at 0.013% and 0.025% of the diet.

Volatile Terpenes

The sweet floral fragrance of cotton, which ranges from pinelike to roselike depending on the cultivar, is at least partly due to the presence of volatile mono- and sesquiterpenes in glanded portions of the plant. Like the terpene aldehyde components found in pigment glands, it is possible that volatile terpenes produced by the plant may have more than one effect on associated insects. The presence of any given compound may be either harmful or beneficial to the plant or to the insect depending on concentration, the biology of the associated organism, and individual circumstance. Individual monoterpenes may also be used as starting materials for the production of more complex compounds which have biological activities of their own.

Occurance and Identification.   An early report of cotton volatile composition by Minyard et al. (44) involved steam distillation of large quantities of leaves and flowers. Major compounds identified included the monoterpenes α-pinene, β-pinene, myrcene, trans-β-ocimene, and limonene (44). Several other monoterpene hydrocarbons were also present in low concentration. Since that report, many other terpenes have been identified in cotton essential oil steam distillates and solvent extracts. These compounds include cyclic hydrocarbons such as bisabolene, caryophyllene, copaene and humulene (45-47), the cyclic epoxide caryophyllene oxide (45), cyclic alcohols such as bisabolol, spathulenol, and the aromatic compound

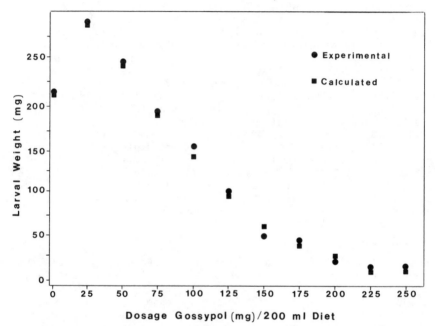

Figure 4. Comparison of mean weights of H. virescens larvae fed
artificial diet containing varying dosages of gossypol with
weights calculated using the exponential/probit model. The LD_{50}
may be determined from the graph.

gossonorol (48-51), and other miscellaneous compounds. Several of these reports are recent and it is likely that more components will be identified in the future. Most of the compounds identified are also known to be present in essential oils of several other species. Representative terpenes of the different classes are shown in Figure 5. The diversity of structure and oxidation state is notable.

Several studies have shown that the volatile chemicals produced by the cotton plant affect the behavior of associated insects. Most studies have concentrated on the feeding and host finding stimulants for the boll weevil Anthonomus grandis (51-60), although a recent study demonstrated that the volatiles also affect the behavior of the beneficial hymenopteran parasitoid, Campoletis sonorensis (50). In all cases studied, a mixture containing several compounds found in the essential oil was more active in eliciting a behavioral response than any single component (51,53,56,58,60), indicating that breeding to reduce quantities of a single component would be unlikely to cause large behavioral changes in associated pest insect behavior. In one instance, an unidentified volatile component which was apparently produced during fractionation at elevated temperatures was shown to repel boll weevils (61). There has been some question as to the relationship between the chemical composition of the essential oil prepared from the plant and the vapor composition above the plant in the field. Volatile collections using cold trapping (62,63) and adsorption on solid support (64) indicate that the air above the plants may be enriched in the more volatile components of the total essential oil.

Several studies have also been carried out to determine qualitative and quantitative differences in volatile chemistry between species and cultivars of cotton and between cotton plants tested during different stages of their growth cycle (62,64-69). The variation between species is useful in taxonomic studies and present work indicates that it should be possible to predict the volatile profiles of at least some lines produced by cross pollination (67).

Comparison of Glanded-Glandless Cotton. We wished to determine if the volatile chemicals play a role in controlling the feeding activity of H. virescens. To our knowledge, this hypothesis has not been previously tested. Since some of the volatile chemicals can produce severe reactions on handling (70-71), it seemed possible that ingestion of reasonably large quantities of such compounds by an insect might also affect the health of the organism. During our previous studies on the behavioral effects of some sesquiterpenes found in ether extracts of G. hirsitum var. Stoneville 213 on the parasitoid C. sonorensis, we had developed procedures for analyzing the volatile terpenoids in small samples of cotton. We first extracted approximately 1.5 g of cotton material in diethyl ether using a micro Soxhlet extractor. The volume was reduced to 0.8 mL by distillation using a short column, the residue was chromatographed on a short florisil column, and the eluate was analyzed by gas chromatography using a 25 m BP1 capillary column. We determined that various plant parts, including flower parts, buds, small bolls, leaves, and roots contained volatile terpenes. Thus volatile terpenes are found in all of the plant parts examined which had pigment glands. Glandless varieties of similar genetic background

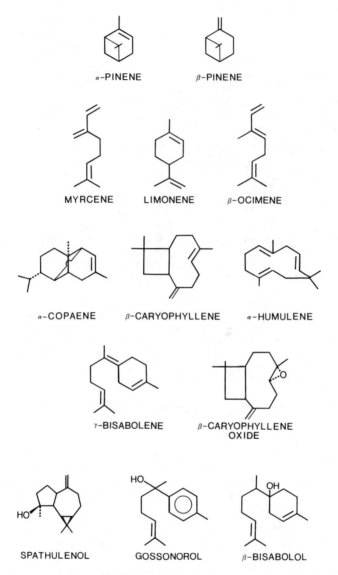

Figure 5.   Representative volatile mono- and sesquiterpenes found in cotton foliar glands.

lacked significant amounts of volatile terpenes (68). It had been determined earlier that the heliocides required myrcene or β-ocimene in their synthesis and were prepared by the plant in ratios which indicated that they were produced by a simple Diels-Alder reaction between these compounds and HGQ (28-31). This indicated that the monoterpenes had to be in high concentration in the pigment glands which contained the heliocides and HGQ, but did not preclude their occurrence in other parts of the plant. We therefore determined their location in the foliar plant parts by piercing 70 cotton flower bud glands with capillary tubes made by drawing glass tubing to a fine point, and removing the gland contents by capillary action. The material in the gland is a dark-colored, mobile liquid, soluble in hexane but insoluble in water. Analysis of this liquid by GC showed that it contained all the volatile terpenes previously found in whole leaf extracts. Extracts of 100 times as much liquid from surrounding tissue or leaf veins had no detectable volatile terpenes. We conclude that the volatile terpenes form the solvent for the terpene aldehydes in the glands (68).

Antibiosis. Since it was known that glandless cotton varieties are more susceptible to damage by H. virescens than the corresponding glanded lines and we have proven that the volatile terpenes were found in the glands, a feeding experiment was performed to determine if a volatile terpene could cause antifeedant activity. The cyclic epoxide caryophyllene oxide (CO) was chosen for the first study due to its rather high concentration in some cultivars and its biological activity in an animal macrophage assay (70). The CO was adsorbed on alphacel and added to prepared diet in the same manner as in the gossypol feeding studies. Neonate larvae fed high dosages of this compound for 7-days were smaller than the controls. A gently sloping S-shaped or sigmoidal growth-response curve was obtained. In view of the high concentrations of volatile terpenes in the cotton plant (68), we believe this resistance mechanism deserves further study. Other cotton terpenes which are known to show antibiosis include pinene, gossonorol, and bisabolol (70,71), and feeding tests utilizing these compounds are planned for the near future.

Synergistic Effects. In view of the antibiotic effect of CO, it was appropriate to investigate the effect of two compounds (G and CO) fed as a mixture. The effect could be simply additive, greater-than-additive (synergistic) or less-than-additive (antagonistic).

We designed an experiment to examine the effect on growth of H. virescens larvae fed G and CO at concentrations thought to be crucial based on the growth-response curves determined in earlier experiments. The G concentrations were chosen to approximate the hormetic maximum (25 mg/200mL), the dose to give a growth approximately equal to the control (50 mg/200mL) and a dose close to the ED_{50} value (100 mg/200 mL; ED_{50} ≈ 118 mg/200 mL). The CO dosages were similarly chosen [growth approximate equal to the control (25 mg/200 mL); and at the approximate ED_{50} (200 mg/200 mL)]. As shown in Table III, the compounds act synergistically with respect to both the hormetic and antibiotic effect. At the hormetic maximum of G (25 mg/200 mL) and the CO dose which gives growth close to equal to the control (75 mg/200 mL), larval weights (177 mg) were 145% of the control weights (122 mg), considerably larger than would be predicted by an additive effect. Furthermore, when CO is mixed at the lowest two levels (25 mg and 75 mg/200 mL) with G at a level slightly below the ED_{50} value (100 mg/200 mL) larval growth is reduced 69% and 74 %, respectively.

Table III. Results of feeding caryophyllene oxide (CO), gossypol (G) or a mixture of G and CO to neonate H. virescens larvae. Larval mean weights are expressed as percent of control.

CO. conc.	G conc. (mg/200 mL diet)			
(mg/200 mL diet)	0	25	50	100
	(Larval Mean Weight as % of Control)[a]			
0	100	133	112	72
25	112	135	120	32
75	84	145	137	21
200	38	113	94	16

[a]Control contained 0 mg of G and CO /200 mL of diet. Absolute range of controls varied from 112 mg to 215 mg in individual experiments.

These results indicate that G and CO may be acting as antifeedants at high dosages and as feeding stimulants at low levels. At low dosages CO appears to act synergistically with low levels of G stimulating feeding and increasing the growth rate (e.g. a larger hormetic effect). Apparently, CO also acts synergistically when G is fed at higher dosages, decreasing feeding and increasing the apparent antibiosis. We will next perform mixed feeding studies with caryophyllene (C) and G since bioassays showed that CO but not C was active in causing release of prostaglandin $F_{2\alpha}$ from alveolar macrophages (70), and in attracting C. sonorensis to plants (50).

Summary· Implications & Future Research

Earlier research on the terpenoid aldehydes demonstrated their antibiotic activity to H. virescens. The authors and their colleagues have established or proposed the following with regard to H. virescens larvae:
1) G, when fed at low levels to larvae, acts as a hormetic agent producing larvae larger than the control.
2) Large variations in larval weight are best explained by grouping larvae according to instar.
3) Simple probit analysis does not reflect the hormetic effect of G, and a new exponential/probit model which more accurately describes this effect has been proposed.
4) Volatile mono- and sesquiterpenes are found almost exclusively in the foliar glands and are not detectable in the foliage of glandless cotton.
5) One of the volatile terpenes, CO, exhibits antibiosis at high levels.

6) When fed as a mixture, **G** and **CO** act synergistically; both hormetic and antibiotic effects are intensified.

7) These findings can be explained if **G** and **CO** are considered as antifeedants at high dosages and as feeding stimulants at low dosages. However, other interpretations are possible.

These results indicate that host plant resistance in cotton will remain a fertile ground for future research. The association of volatile terpenes with glands, and the hormetic and antibiotic effects of one member of this group, are persuasive arguments for bioassaying other compounds. The unexpected synergistic effects observed when **G** and **CO** are fed as a mixture demand further study. Thus we intend to expand these studies to more fully define the contours of this growth-response surface, and extend this research to other terpenes.

Work at our laboratory indicates that the quality and quantity of the volatile terpenes can be genetically manipulated (72). For example, five commercial Upland Cottons (G. hirsutum) [CAMD-E, PAY-202, STO-213, LANK-57, and ACALA SJ-1] all contain between 23-33% caryophyllene (**C**) but only 0.3 to 0.7% **CO** (67). Preliminary results indicate that in some G. arboreum lines, **C** and **CO** comprise up to 25% of the volatile terpenes. We therefore plan to assay **C** to determine its activity against H. virescens. Since **C** was inactive in animal cell studies, it may have no effect on feeding by H. virescens larvae. In that case it would be beneficial to increase the **CO/C** ratio in the plant, probably by a single gene transfer. Before such **CO/C** ratio manipulation were recommended, some **G/CO/C** ternary bioassay data will be required.

These results should be a valuable tool in assisting the cotton breeders in their efforts to improve host plant resistance. In the past it has been the breeder and geneticist who have led in developing new resistant lines. After these lines proved their worth in field tests, the chemist has been invited to determine the involvement of allelochemics. This process will remain the norm, but experiments such as ours may allow predictions regarding combinations of chemicals which would be most efficacious in protecting the plant. This information combined with knowledge concerning genetic control of allelochemic production will allow plant breeders to engineer superior plants in less time.

Acknowledgments

The authors thank G. W. Elzen, A. A. Bell, and S. B. Vinson for their contributions to this work, and Deborah Begin, Patricia Darnell and Jose Fourquet for their excellent technical assistance.

This manuscript approved as TAES No. 20779 by the Director of the Texas Agricultural Experiment Station.

Mention of firm names or trade products does not imply endorsement by the U.S. Department of Agriculture over other firms or similar products not mentioned.

Literature Cited

1. Lukefahr, M. J.; Fryxell, P. A. Econ. Bot. 1967, 21, 128.
2. Adams, R.; Geissman, T. A. Chem. Rev. 1960, 60, 555.
3. Berardi, L. C.; Goldblatt, L. A. in "Toxic Constituents of Plant Foodstuff"; Liener, I. E., Ed.; Academic Press: New York, 1980, p. 183.
4. Abou-Donia, M. B. Residue Review 1976, 61, 125.
5. National Coordinating Group on Male Antifertility Agents (China) Gynecol. Obstet. Invest. 1979, 10, 163.
6. Manmade, A.; Herlihy, P.; Quick, J.; Duffley, R. P.; Burgos, M.; Hoffer, A. P.; Experientia 1983, 39, 1276.
7. Murthy, R. S. R.; Basu, D. K. Current Science 1981, 50, 64.
8. Zatuchini, G. E.; Osborn, C. K. Res. Frontiers Fertil. Regul. 1981, 1, 1.
9. Abou-Donia, M. B.; Pieckert, J. W. Life Sci. 1974, 14, 1955.
10. Tso, W. W.; Lee, C.-S.; Tso, M. Y. W. Arch. Androl. 1982, 9, 31.
11. Adeyemo, O.; Chang, C. Y.; Segal, S. J.; Koide, S. S. Arch. Androl. 1982, 9, 343.
12. Haspel, H. C..; Corin, R. E.; Sonenberg, M. Fed. Proc. 1982, 41, 671.
13. Reyes, J.; Allen, J.; Tanphaichitr, N.; Bellve', A. R.; Benos, D. J. J. Biolog. Chem. 1984, 259, 9607.
14. Altschul, A. M.; Lyman, C. M.; Thurber, F. H. in "Processed Plant Protein Food Stuffs", Altschul, A. M. Ed. Academic Press: New York, 1958, p. 497.
15. Mace, M. E.; Bell, A. A.; Stipanovic, R. D. Phytopath. 1974, 64, 1297.
16. Hunter, R. E.; Halloin, J. M.; Veech, J. A.; Carter, W. W. Plant and Soil 1978, 50, 237.
17. Bell, A. A. in "Cotton Physiology: A Treatise" Stuart, J. M.; Mauney, J. R. Eds.; Cotton Foundation, Memphis, TN, 1985, in press; see also, Bell, A. A.; Mace, M. E.; Stipanovic, R. D. this book.
18. Bell, A. A.; Stipanovic, R. D.; Howell, C. R.; Fryxell, P. A. Phytochem. 1975, 14, 225.
19. Stipanovic, R. D.; Bell, A. A.; Howell, C. R. Phytochem. 1975, 14, 1809.
20. Stipanovic, R. D.; Bell, A. A.; Mace, M. E.; Howell, C. R. Phytochem. 1975, 14, 1077.
21. Veech, J. A.; Stipanovic, R. D.; Bell, A. A. J. Chem. Soc., Chem. Comm. 1976, 144.
22. Stanford, E. E.; Viehoever, A. J. Agr. Res. 1918, 13, 419.
23. Chan, B. G.; Waiss, A. C., Jr. Beltwide Cotton Prod. Res. Conf. Proc. 1981, 49.
24. Hedin, P. A.; Jenkins, J. W.; Collum, D. H.; White, W. H.; Parrott, W. L.; MacGown, M. W. Experientia 1983, 39, 799.
25. Elliger, C. A.; Chan, B. G.; Waiss, A. C., Jr. J. Econ. Entomol. 1978, 71, 161.
26. Stipanovic, R. D.; Bell, A. A.; Lukefahr, M. J. in "Host Plant Resistance to Pests", ACS Symposium Series, No. 62; Hedin, P. A., Ed.; American Chemical Society: Washington, D. C., 1977, p. 197.

27. Gray, J. R.; Mabry, T. J.; Bell, A. A.; Stipanovic, R. D.; Lukefahr, M. J. J. Chem. Soc., Chem. Comm. 1976, 109.
28. Stipanovic, R. D.; Bell, A. A.; O'Brien, D. H.; Lukefahr, M. J. J. Agr. and Food Chem. 1978, 26, 115.
29. Stipanovic, R. D.; Bell, A. A.; O'Brien, D. H.; Lukefahr, M. J. Tetrahedron Lett., 1977, 567.
30. Stipanovic, R. D.; Bell, A. A.; O'Brien, D. H.; Lukefahr, M. J. Phytochem. 1978, 17, 151.
31. Bell, A. A.; Stipanovic, R. D.; O'Brien, D. H.; Fryxell, P. A. Phytochem. 1978, 17, 1297.
32. Bell, A. A. Personal communication.
33. Stipanovic, R. D.; Bell, A. A.; O'Brien, D. H. Phytochem. 1980, 19, 1735.
34. McMichael, S. C. Agr. J. 1959, 51, 30.
35. Bottger, G. T.; Sheehan, E. T.; Lukefahr, M. J. J. Econ. Entomol. 1964, 57, 283.
36. Lukefahr, M.J.; Martin, D. F. J. Econ. Entomol. 1966, 176.
37. Jenkins, J. N.; Maxwell, F. G.; Lafever, H. N. J. Econ. Entomol. 1966, 59, 352.
38. Chan, B. G.; Waiss, A. C., Jr.; Binder, R. G.; Elliger, C. A. Entomologia Exp. App. 1978, 24, 94.
39. Stebbing, A. R. D. The Sci. Total Environ. 1982, 22, 213.
40. Clark, K.U. Proc. Roy. Ent. Soc. Lond. A. 1957, 32, 35.
41. Montalva, Roxanne, Private Communication.
42. Finney, J. D. in "Probit Analysis," Cambridge Univ. Press. Cambridge, England, 1952.
43. Williams, H. J.; Stipanovic, R. D.; Smith, L. A. In prep
44. Minyard, J. P.; Tumlinson, J. H.; Hedin, P. A.; Thompson, A. C. J. Agr. Food Chem. 1965, 13, 599.
45. Minyard, J. P.; Tumlinson, J. H.; Thompson, A. C.; Hedin, P. A. J. Agr. Food Chem. 1966, 4, 332.
46. Hedin, P. A.; Thompson, A. C.; Gueldner, R. C.; Ruth, J. M. Phytochem. 1972, 11, 2118.
47. Minyard, J. P.; Tumlinson, J. H.; Thompson, A. C.; Hedin, P. A. J. Agr. Food Chem. 1967, 15, 517.
48. Minyard, J. P.; Thompson, A. C.; Hedin, P. A. J. Org. Chem. 1968, 33, 909.
49. Hedin, P. A.; Thompson, A. C.; Gueldner, R. C. Phytochem. 1971, 10, 1693.
50. Hedin, P. A.; Thompson, A. C.; Gueldner, R. C.; Minyard, J. P. Phytochem. 1971, 10, 3316.
51. Elzen, G. W.; Williams, H. J.; Vinson, S. B. J. Chem. Ecol. 1984, 10, 1251.
52. Hanny, B. W.; Thompson, A. C.; Gueldner, R. C.; Hedin, P. A. J. Agr. Food Chem., 1973, 21, 1004.
53. Gueldner. R. C.; Thompson, A. C.; Hardee, D. D.; Hedin, P. A. J. Econ. Entomol. 1970, 63, 1819.
54. McKibben, G. H.; Mitchell, E. B.; Scott, W. P.; Hedin, P. A. Environ. Entomol. 1977, 6, 804.
55. Keller, J. C.; Maxwell, F. G.; Jenkins, J. N.; Davich, T. B. J. Econ. Entomol. 1963, 56, 110.
56. Hedin, P. A.; Miles, L. R.; Thompson, A. C.; Minyard, J. P. J. Agr. Food Chem. 1968, 16, 505.

57. McKibben, G. H.; Hedin, P. A.; McLaughlin, R. E.; Davich, T.
    B. J. Econ. Entomol. 1971, 64, 1493.
58. Minyard, J. P.;, Hardee, D. D.; Gueldner, R. C.; Thompson, A.
    C.; Wiygul, G.; Hedin, P. A. J. Agr. Food Chem. 1969, 17,
    1093.
59. Thompson, A. C.; Wright, B. J.; Hardee, D. D.; Gueldner, R.
    C.; Hedin, P. A. J. Econ. Entomol. 1970, 63, 751.
60. Gueldner, R. C.; Thompson, A. C.; Hardee, C. C.; Hedin, P. A.
    J. Econ. Entomol. 1970, 63, 1819.
61. Maxwell, F. G.; Jenkins, J. N.; Keller, J. C. J. Econ.
    Entomol. 1963, 56, 894.
62. Thompson. A. C.; Baker, D. N.; Gueldner. R. C.; Hedin, P. A.
    Plant Physiol. 1971, 48, 50.
63. Keller, J. C.; Maxwell, F. G.; Jenkins, J. N.; Davich, T. B.
    J. Econ. Entomol. 1963, 56, 110.
64. Hedin, P. A. Environ. Entomol. 1976, 5, 1234.
65. Hedin, P. A.; Thompson, A. C.; Gueldner, R. C.; Minyard, J. P.
    Phytochem. 1972, 11, 2356.
66. Kumamoto, J.; Waines, J. G.; Hollenberg, J. L.; Scora, R. W.
    J. Agr. Food Chem. 1979, 27, 203.
67. Elzen, G. W.; Williams H. J.; Bell, A. A.; Stipanovic, R. D.;
    Vinson, S. B., Ninth Cotton Dust Research Conf. New Orleans,
    LA, Jan. 9-11, 1985, p. 47.
68. Elzen, G. W.; Williams, H. J.; Bell, A. A.; Stipanovic, R. D.;
    Vinson, S. B. J. Agr. Food Chem. Submitted.
69. Thompson, A. C.; Hanny, B. W.; Hedin, P. A.; Gueldner, R. C.
    Amer. J. Bot. 1971, 58, 803.
70. Ziprin, B. L.; Beerwinkle, K. R.; Williams, H. J. Proceed.
    Ninth Cotton Dust Research Conf., New Orleans, LA, Jan. 9-11,
    1985, p. 120.
71. Gleason, M. N.; Gosselin, R. E.; Hodge, H. C. "Clinical
    Toxicology of Commercial Products"; Williams & Wilkins:
    Baltimore, MD, 1957, p. 82.
72. Bell, A. A. Personal Communication.

RECEIVED August 9, 1985

3-Nitropropionate in Crownvetch: A Natural Deterrent to Insects?

R. A. Byers, David L. Gustine, B. G. Moyer, and D. L. Bierlein

U.S. Regional Pasture Research Laboratory, USDA, Agricultural Research Service, University Park, PA 16802

3-Nitropropionic acid (NPA) and two glucose esters, cibarian (diester) and karakin (triester), added to a pinto bean diet and fed to Sparganothis fruitworms, Sparganothis sulfureana (Clemens), significantly increased mortality of larvae and reduced pupal weight of both males and females over four generations. These compounds isolated from leaves, flowers, and stems of crownvetch, Coronilla varia L., are also toxic to nonruminant animals (chickens and pigs) and to the cabbage looper, Trichoplusia ni Hübner. Cabbage loopers lose weight and eventually die when fed crownvetch leaves. However, Sparganothis fruitworms feed and develop normally on crownvetch, but in greater numbers in autumn, presumably because the lower temperatures reduce NPA concentrations, which in turn permits increased survival of insects. Attempts to reduce summertime peak NPA concentrations in crownvetch by selective breeding may produce an insect susceptible cultivar.

Toxic Effects of 3-Nitropropionic Acid (NPA)

Allomones, allelochemicals deleterious to the receiving organism, have been implicated in plant defense against herbivores (1). Rhoades (2) credits Stahl (3) in 1888 as the first person to suggest that some chemicals in plants may have evolved as protection against herbivores. Rhoades (2) suggested that two main lines of evidence have been used to show antiherbivore activity of allelochemicals: one involves deterrency, the other antibiosis. Preference of herbivores for a series of tissues is compared with the concentration of allelochemicals in these tissues. If the herbivore is deterred, a negative correlation is the result. Deterrency can be corroborated by extraction of the substance and incorporation into artificial diet. Antibiosis is measured by some fitness measure such as growth rate, fecundity, number of viable

progeny when herbivores are fed various tissues or artificial diets
containing these allelochemicals.

Hedin (4) listed many allelochemicals from plants that
influence insect behavior and development. Some of these compounds
are important in plant defense acting as feeding deterrents,
repellents, or toxins to insects. Unfortunately, some plants with
defensive allomones may also be forages for domestic animals. Such
is the case for crownvetch, Coronilla varia L., a legume with high
protein content (5) that provides satisfactory forage for ruminant
animals (6) but is toxic to nonruminant animals such as chickens
and pigs (7). Gustine et al. (8) proved that the toxicity was
caused by NPA in the forage. NPA is also toxic to at least one
insect. We showed that NPA added to a pinto bean diet adversely
affected development of cabbage loopers, Trichoplusia ni Hübner
(9). Gustine et al. (8) and Moyer et al. (10) showed NPA exists in
the plant as glucose esters. We found that NPA and the NPA esters
cibarian (diester) and a mixture of coronillin and karakin
(triesters), significantly increased mortality and retarded the
development of first instar larvae of cabbage loopers fed
artificial diets containing these compounds (Figures 1 and 2).
First instar larvae that died showed little to no evidence of
feeding on diets, implying NPA and its esters may act as feeding
deterrents. However, these chemicals probably also act as toxins
because of dramatic reductions in pupal development and pupal
weight and higher mortality caused by increasing concentrations of
NPA and its triesters in the diet (Table I). Furthermore, although
NPA and its esters had no detectable effect on longevity of adults,
every compound tested significantly lowered fecundity (Figure 3).

Kendall et al. (11) observed NPA acted as a feeding deterrent
to meadow voles Microtus pennsylvanicus. They fed adult male voles
crownvetch leaves grown under different nitrogen regimes or
synthetic diets containing various amounts of NPA mixed with potato
starch. Concentration of NPA in leaves was from 0.7 to 1.7% and
concentration in the diet was from 0 to 2.5%. They found meadow
vole intake followed a quadratic response curve with an inverse
relationship between NPA and intake. Diets containing more than 1%
NPA were almost completely rejected.

Therefore, NPA may be an allomone protecting crownvetch acting
as a deterrent to herbivores. Meadow voles are certainly
considered herbivores that could encounter crownvetch in the field.
However, cabbage loopers may not adequately represent insect
herbivores on crownvetch because they use Cruciferae and other
plant families as host plants rather than Leguminosae.
Consequently, we collected several insect species feeding on
crownvetch in late summer to study the effects of NPA on these
herbivores. A leaf tier, the Sparganothis fruitworm, Sparganothis
sulfureana (Clemens), produced eggs from a field collected moth and
could be reared on pinto bean diet.

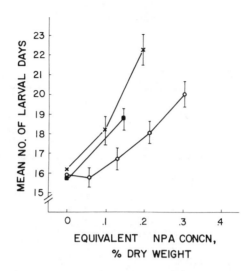

Fig. 1. Effect of NPA, cibarian, and a mixture (karakin and coronillin) on the larval period of cabbage loopers. X- NPA, Solid Square-Mixture, O-Cibarian. Vertical lines represent the 95% confidence interval. Reproduced with permission from Ref. 9. Copyright 1977, "Entomological Society of America."

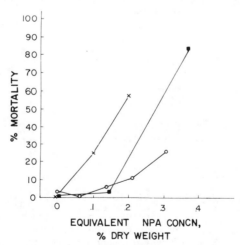

Fig. 2. Effect of NPA, cibarian, and a mixture (karakin and coronillin) on mortality of first instars of cabbage loopers, X-NPA, Solid Square-Mixture, O-Cibarian. Vertical lines represent the 95% confidence interval. Reproduced with permission from Ref. 9. Copyright 1977, "Entomological Society of America."

Table I.   Effect of 3-Nitropropionic Acid (NPA) and Its Diesters
           Cibarian and a Mixture of Triesters, Coronillin,
           and Karakin on Pupae of Cabbage Looper,
           Trichoplusia ni

Conc. NPA Equivalents % Dry Wt.	Mean Days as Pupa	Mean Pupal Wt. (mg)	Mortality (%)
	NPA		
0	9.1 a	254.6 de	0
0.1	9.6 cd	261.8 e	0
0.125	10.1 ef	235.4 c	0
0.2	9.0 a	170.7 a	8.6
0.5	--	--	100.6
	Cibarian		
0	9.5 bc	233.0 c	2.7
0.062	9.9 de	242.6 cd	0
0.078	9.5 bc	233.6 c	0
0.128	9.7 cd	241.7 cd	2.8
0.217	9.2 a	201.4 b	5.6
0.310	9.4 abc	177.5 a	14.7
	Coronillin and Karakin		
0	9.6 cd	255.2 de	2.7
0.148	10.2 f	204.9 b	2.7
0.310	--	--	100.0

Means with the same letter in a column are not significantly
different at P=0.01. Duncan's Multiple Range Test.

Rearing Sparganothis Fruitworms

A female moth, collected from a roadside bank of crownvetch in late
summer, laid eggs on Parafilm and on the sides and lid of a plastic
container in the laboratory. The eggs were incubated on moist
filter paper at 20-25 C with a 14-hour photophase. Newly emerged
larvae were transferred to artificial diet in plastic cups (27 cc),
one larva per cup. The basic diet consisted of 80 g dry ground
pinto beans, 12 g Torula yeast, 0.8 g methyl paraben, 0.4 g sorbic
acid, and 1.2 g ascorbic acid. Three compounds, NPA, cibarian, and
karakin were added as powders to the basic diet. All ingredients
were added to 124 ml tap water and 0.7 ml formaldehyde in a
blender. Melted agar (4.5 g in 100 ml tap water) was added, and
the mixture was blended for 15 sec. The diet was poured into petri
dishes, cooled to room temperature, and either cut into pieces
(8 cc) and placed in plastic cups or stored at 5 C until needed.
Larvae were reared to adults in an incubator at 22-25 C and 14-hour
photophase. Pupae were weighed the day following pupation and
their sex recorded. Pairs of moths from each treatment were fed 4%
honey-water while confined to plastic containers (550cc).

Each container was provided with strips of folded Parafilm for oviposition sites. The number of eggs/pair was recorded.

Insects were reared through four generations but the culture was lost because of infertility of fourth generation eggs. Treatments, on a percent of dry weight basis, varied for each generation and were expressed as percent by dry weight of NPA or as the NPA equivalent of the esters. Treatments in the first generation were: NPA (0.15 and 0.30%) cibarian (0.09 and 0.18%) and karakin (0.11 and 0.22%). There were three control treatments with 18 replications of each treatment. Treatments and replications for generations 2 and 3 were identical to generation 1 except there were only two controls in the second generation and only one in the third. Treatments for the fourth generation were: NPA 0.3%, cibarian 0.3% and karakin 0.33%, and an untreated control with 18 replications. Data were pooled for the four generations and analysed by analysis of variance. Treatment means were separated by the Waller-Duncan Bayesian k ratio t-test. Linear regressions were computed for larval mortality, longevity of larvae, pupae, and adults, and male and female pupal weights versus concentration of NPA or NPA equivalents for the esters.

Effect of NPA on Larvae of Sparganothis Fruitworms

Increasing concentrations of NPA increased mortality of larvae of S. sulfureana. The regression equation for NPA concentration versus percent mortality for larvae was:

$$\hat{Y} = 7.9 + 65.2X$$

where X = NPA concentration percent dry weight. The value of F ratio was 5.46 (P = 0.04) and R^2 = 0.38. Most of the mortality occurred during the first instar, with little to no production of frass pellets and scant evidence of feeding.

NPA also retarded larval development of females because the mean number of days spent as a larva increased significantly at increasing NPA concentrations. The regression equation for NPA concentration versus number of days as a larva for females was:

$$\hat{Y} = 25.7 + 24.0X$$

where X = NPA concentration percent dry weight. The value of the F ratio was 14.57 (P = 0.0003) and R^2 = 0.16. However, increasing concentrations of cibarian and karakin did not significantly slow larval development. None of the compounds retarded larval development of males. On the contrary, karakin significantly shortened the longevity of male larvae (Table II).

Effect of NPA on Pupae of Sparganothis Fruitworms

No significant differences between treatments were detected for pupal mortality or mean days as pupa. However, pupal weights of both sexes were significantly reduced below weights of controls by increasing concentrations of NPA and both esters (Figures 4 and 5).

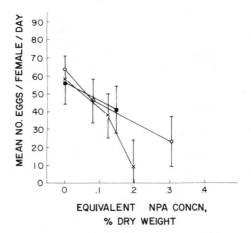

Fig. 3. Effect of NPA, cibarian, and a mixture (karakin and coronilliin) on fecundity of cabbage loopers. X-NPA, Solid Square-Mixture, O-Cibarian. Vertical lines represent the 95% confidence interval. Reproduced with permission from Ref. 9. Copyright 1977, "Entomological Society of America."

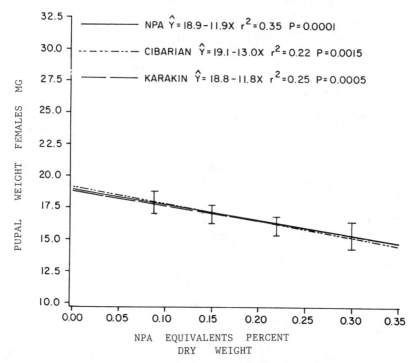

Fig. 4. Effect of NPA, cibarian, and karakin on pupal weight of male Sparganothis fruitworms. Vertical lines represent the 95% confidence interval.

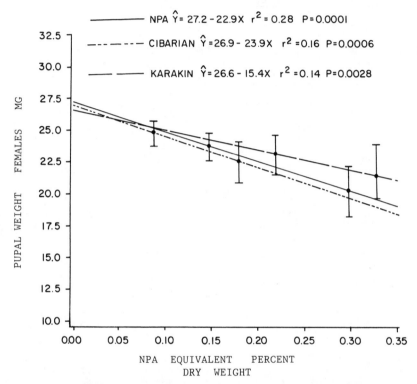

Fig. 5.   Effect of NPA, cibarian, and karakin on pupal weight of female Sparganothis fruitworms.   Vertical lines represent the 95% confidence interval.

Table II. Effect of 3-Nitropropionic Acid (NPA) and Its Esters
           Cibarian and Karakin on Larvae of
           Sparganothis sulfureana

Treatment NPA Conc. % Dry Wt.	Mean % Mortality	Mean Days as Larva	
		Females	Males
		NPA	
0	6.9	25.7	27.1
0.15	20.4	29.3	28.3
0.30	27.8	32.9	26.3
	18.8 a	28.7 a	27.1 a
		Cibarian	
0	5.6	27.8	26.4
0.09	13.0	27.3	26.6
0.18	7.4	26.6	26.9
0.30	27.8	29.2	26.6
	11.2 a	27.4 a	26.6 a
		Karakin	
0	5.6	26.8	23.1
0.11	25.9	28.3	25.0
0.22	13.0	27.0	26.8
0.33	22.2	28.0	25.2
	18.1 a	27.5 a	25.5 b

Means with the same letter in a column are not significantly
different. k=100 (Waller-Duncan Bayesian k ratio t-test).

Effect of NPA on Adults of Sparganothis Fruitworms

Increasing concentrations of NPA in the larval diet had little
effect on female longevity. However, increasing concentrations of
cibarian in the larval diet decreased the longevity of females.
The regression equation was:

$$\hat{Y} = 15.3 - 23.2X$$

where X was the concentration of cibarian in NPA equivalents by dry
weight in the larval diet. The F ratio was 5.58 (P = 0.02) and
$R^2 = 0.07$. Female moths lived significantly longer when larvae
were fed cibarian rather than karakin (Table III). Male moths
lived longer with increasing concentrations of karakin in the
larval diet. The regression equation was:

$$\hat{Y} = 1.97 + 29.4X$$

where X was concentration of karakin in NPA equivalents by dry
weight. The F ratio was 10.05 (P = 0.0025) and $R^2 = 0.15$. The sex
ratio favored females at higher concentrations of NPA and cibarian
but not karakin. Females lived longer than males. No significant
differences were detected among treatments for mean number of eggs
laid or eggs/female/day.

Table III.   Effect of 3-Nitropropionic Acid (NPA) and Its Esters
Cibarian and Karakin on Adults of
Sparganothis sulfureana

Treatment NPA Conc. % Dry Wt.	Mean Length of Adult Life		Sex Ratio
	Females	Males	
	NPA		
0	11.7	6.2	1:1
0.15	12.7	4.9	1:1.3
0.30	9.3	7.9	1.4:1
	11.3 ab	6.5 a	
	Cibarian		
0	13.5	6.4	1:1.8
0.09	15.4	7.9	1:1.2
0.18	10.9	3.2	1.4:1
0.30	5.4	14.4	1.6:1
	12.9 a	6.4 a	
	Karakin		
0	10.1	8.4	1.4:1
0.11	9.9	5.2	1.2:1
0.22	8.3	8.7	1:1
0.33	11.2	11.0	1:1.3
	9.1 b	6.7 a	
Mean of Sexes	11.1	6.4	

F ratio for differences between sexes = 8.95, Probability = 0.01.

Means with the same letter in a column are not significantly
different at P=0.05.   Waller-Duncan Bayesian k ratio t-test.

NPA as an Allomone in Crownvetch

NPA and its esters adversely affected growth and development of
larvae and reduced pupal weight of Sparganothis fruitworms, but to
a lesser degree than for the cabbage looper. This was not
unexpected since crownvetch is not a host for the cabbage looper,
but Sparganothis fruitworms can be reared to adults on crownvetch.
However, the larval mortality and reduced pupal weights recorded
for both insects indicates NPA may be a toxic allomone.
    Feeny (12-13) proposed the term "quantitative" defenses of
plants for chemicals of low toxicity, such as tannins, constituting
up to 60% of the dry weight. Qualitative defenses include
allomones of high toxicity but usually present in low
concentrations in plants (14). Most qualitative defenses occur in
ephemeral or unapparent plants (annuals and short-lived perennials
which are widely dispersed) and are designed to interfere with
internal metabolism of herbivores, whereas most predictable or
apparent plants (trees) rely on quantitative defenses such as
tannins designed to reduce digestibility (2). NPA qualifies as a

qualitative defensive chemical because it constitutes only
0.25-1.25% fresh weight of crownvetch leaves and reaches maximum
concentration during flowering, decreasing to very low levels after
blooming and seed pod formation (10-15-16).

Rhoades (2) postulated that flower nectar with toxic
constituents serves to maximize flower constancy in pollinators.
All plants pollinated by animal vectors benefit from
flower-constant behavior since nectar removed by visitors not
carrying conspecific pollen is wasted. However, rare or widely
dispersed plants should be more strongly selected for adaptations
that maximize pollinator constancy than should common or clumped
plants (2). Perhaps in crownvetch there has been selection for
constancy in flower visitors using high concentrations of NPA to
exclude all but specific pollinators.

The ability of insects to attack crownvetch (17) does not
contradict our hypothesis that NPA is a deterrent to insects. Even
though NPA in diets adversely affects Sparganothis fruitworm, this
insect can use crownvetch as a host plant. However, we observed
this insect along highway roadside banks more frequently in
September and October when lower temperatures would be expected to
reduce NPA concentrations in the plant (16). Other insects have
been observed attacking crownvetch in autumn. The moth,
Stomopteryx palpilineela (Chambers) defoliated crownvetch on a
roadside bank in October near Newport, Perry Co., PA (18). We have
collected redbanded leafrollers, Argyrotaenia velutinana (Walker)
feeding on crownvetch near State College, Centre Co., PA, in
November. The increased ability of these insects to attack
crownvetch in autumn may be a consequence of the low levels of NPA
because these insects have lower populations on crownvetch earlier
in the season when NPA concentrations would be expected to be much
higher.

Plant breeding programs to reduce NPA levels in crownvetch
because of undesirable side effects to nonruminant animals may
possibly produce an insect-susceptible cultivar. Rhoades (19)
concluded that all plant breeding programs designed to control
specific pests or improve forage quality to domestic animals by
reducing the concentrations of secondary metabolites in plants
incur high risk of increased general pest problems. He cited
several examples: plants selected for low levels of curcurbitacin
to control cucumber specific beetles, are decimated by the
twospotted spider mite; glandless cotton low in gossypol developed
to control boll weevil is more susceptible to blister beetles;
strains of alfalfa low in saponin content to control alfalfa weevil
and bloat in cattle are more heavily attacked by pea aphids; and
lupine cultivars selected for low alkaloid content, to improve
palatability and nutritional value to cattle, are heavily attacked
by thrips. Therefore, selective breeding for lower levels of NPA
in crownvetch to improve its value as a feed for nonruminants may
not be wise because insects may be better able to utilize
crownvetch as a host plant. Consequently, such selections of
crownvetch with low levels of NPA should be screened for
susceptibilty to insect damage before release to breeders.

Literature Cited

1. Whittaker, R.H.; Feeny, P.P. Science 1971, 171, 757-70.
2. Rhoades, D.F. In "Herbivores, Their Interaction with Secondary Plant Metabolites"; Rosenthal, G.A.; Janzen, D.H., Eds.; Academic Press: New York, 1979; pp. 3-54.
3. Stahl, E.; Jena, Z. Med. Naturwiss. 1888, 22, 557-684.
4. Hedin, P.A.; Jenkins, J.A.; Maxwell, F.G. In "Host Plant Resistance to Pests"; Hedin, P.A., Ed.; ACS SYMPOSIUM SERIES No. 62, American Chemical Society: Washington, D.C., 1977; pp. 231-75.
5. Shenk, J.S.; Risius, M.L. Agron. J. 1974, 66, 386-9.
6. Peterson, A.D.; Baumgardt, B.R.; Long, T.A. J. Anim. Sci. 1974, 38, 172-7.
7. Shenk, J.S.; Wangsness, P.J.; Leach, R.M.; Gustine, D.L.; Gobble, J.L.; Barnes, R.F. J. Anim. Sci. 1976, 42, 617-21.
8. Gustine, D.L.; Shenk, J.S.; Moyer, B.G.; Barnes, R.F. Agron. J. 1974, 66, 636-9.
9. Byers, R.A.; Gustine, D.L.; Moyer, B.G. Environ. Entomol. 1977, 6, 229-32.
10. Moyer, B.G.; Pfeffer, P.E.; Moniot, J.L.; Shamma, M.; Gustine, D.L. Phytochemistry 1977, 16, 375-7.
11. Kendall, W.A.; Hill, R.R. Jr.; Shenk, J.S. Agron. J. 1979, 71, 613-6.
12. Feeny, P.P. In "Coevolution of Animals and Plants"; Gilbert, L.E.; Raven, P.H., Eds.; Univ. Texas Press: Austin, 1975, pp. 1-19.
13. Feeny, P.P. Recent Adv. in Phytochem. 1976, 10, 1-40.
14. Strong, D.R.; Lawton, J.H.,; Southwood, R. In "Insects on Plants, Community Patterns and Mechanisms"; Harvard Univ. Press: Cambridge, 1984, p. 30.
15. Gustine, D.L. Crop Sci. 1979, 19, 197-203.
16. Faix, J.J.; Gustine, D.L.; Wright, M.J. Agron. J. 1978, 70, 689-91.
17. Wheeler, A.G., Jr. Can. Entomol. 1974, 106, 897-908.
18. Valley, K.; Wheeler, A.G., Jr. Ann. Entomol. Soc. Am. 1976, 69, 317-24.
19. Rhoades, D.F. "Variable Plants and Herbivores in Natural and Managed Systems"; Denno, R.F.; McClure, M.S., Eds.; Academic Press: New York, 1983, p. 172.

RECEIVED October 9, 1985

9

Between-Year Population Variation in Resistance of Douglas-fir to the Western Spruce Budworm

Rex G. Cates[1] and Richard A. Redak[2]

[1] Department of Botany and Range Science, Brigham Young University, Provo, UT 84602
[2] Department of Entomology, Colorado State University, Fort Collins, CO 80523

Patterns of interactions among individuals within a population of Douglas-fir (Pseudotsuga menziesii var. glauca [Beisn.] Franco) and the western spruce budworm (Choristoneura occidentialis Freeman) were monitored to determine relationships of foliage and physical tree variables to budworm success. Terpene chemistry of the current year's foliage was an important factor in reducing female and male budworm dry weight production. In 1981, trees with high beta-pinene produced smaller females, while in 1982, trees with high bornyl acetate and an unidentified sesquiterpene produced the smaller females. For males, limonene and carene were inversely related to male dry weight production in 1981, while bornyl acetate and terpene evenness were inversely related to male biomass production in 1982. Trees determined to be relatively resistant in 1981 remained resistant in 1982. Reasons for the shift in production of certain terpenes between years has not been resolved but foraging by the insect may be implicated.

In a continuing effort to elucidate the patterns of interaction among individuals within a population of Douglas-fir (Pseudotsuga menziesii var. glauca [Beisn.] Franco) and the western spruce budworm (Choristoneura occidentialis Freeman), we have monitored foliage and physical tree variables to determine their relationship to budworm success (1,2,3). The general objective of the research was to simultaneously study the possible effects of foliage quality (as defined by foliar nitrogen and monoterpene content), host tissue phenology, and tree physical characteristics upon larval growth and adult biomass production which, in the budworm, is highly correlated with egg production. Specifically, the study was designed to determine (1) the extent to which individuals within a population of host plants exhibit herbivore resistance as bioassayed by budworm dry weight produc-

0097-6156/86/0296-0106$06.00/0
© 1986 American Chemical Society

tion, (2) which of the foliage quality, tissue phenology, and
tree physical characteristics were correlated with this
resistance, (3) if the parameters associated with resistance vary
from year to year among the same individuals in a single population of Douglas-fir, and (4) if differences exist between budworm
sexes with regard to resistance factors.

Methods and Materials

Field methods. The research site is located in Barley Canyon,
Santa Fe National Forest, approximately 2.4 km northeast of
Fenton Lake, New Mexico, U.S.A. Details of the site and selection of sample trees used in 1981 and 1982 studies are outlined
(3,4). Age, height, dbh, live crown diameter, tree bole radius
(as determined from increment cores), 5-year growth of annual
rings, and crown ratio (crown height divided by tree height) were
determined and used as vigor indices for each sample tree. After
foliage had ceased growth in early August, 1981, average inter-
nodal growth for 1981 was determined by taking the average of
three internode lengths from three branches in the north-facing,
midcrown section of the sample trees. In early August, 1982,
the same procedure was used to determine the 1982 internode
length. Average internode lengths were used as estimates of the
amount of tissue available to budworm for 1981 and 1982. Time at
which each tree began to flush tissue (budbreak) was recorded as
days since 1 June 1981 or 1982.

Budworm larvae used in the study were collected about mid-
June each year from a vigorously growing population that was
infesting Douglas-fir within the Carson National Forest near
Taos, New Mexico. Branches from infested trees were cut, placed
in insulated, air conditioned trucks, and transported to an
environmental chamber (22°C, 35% rh) housed in the Biology
department, University of New Mexico, Albuquerque, N.M. Within
12 h of collection, larvae were removed from the infested
branches, sorted to third instar, placed into plastic vials with
perforated lids, and returned to the environmental chamber to
minimize larval stress. Each vial contained a sprig of young
Douglas-fir tissue.

Within a day of collection, experimental animals were
transported to the Barley Canyon site and placed on sample trees.
Ten larvae were placed in a nylon screen bag enclosing one
branch which contained at least 10 new foliage buds. Any resi-
dent larvae present on the branches were removed prior to
bagging. Five bags, each with 10 larvae, were placed on each
tree yielding a total of 50 larvae per tree. All bags were
placed within the north-facing, midcrown quarter of each tree.
The nylon bags used, while preventing larval escape, allowed 70%
of full sunlight to pass through. Thus, reduction in photosynthesis due to bagging was not expected.

Experimental larvae were allowed to feed on sample trees
until pupation. All larvae pupated within about three weeks
after placement on the trees. Near mid-July, bags were cut from
the trees and transported to the laboratory. Pupae were removed
from the bags, placed in labeled paper cups, and sealed with

cheesecloth. Cups were placed in an environmental chamber (22°C, 35% rh, 18:6 h light;dark photoperiod), and adults were allowed to emerge. Of the adults that emerged, approximately 95% did so within 72 h of one another indicating similar developmental rates among budworms reared on different trees. Upon emergence, all adults were killed, sexed, and weighed. Animals were then dried for 72 h at 60°C. Dry weights subsequently were determined.

Chemical methods. Approximately 5g fresh weight of current year's foliage were collected from each tree when the larvae were in the fifth and sixth instars. All foliage was collected within the same midcrown quarter of the trees where experimental insects were placed. Collected tissue was put in labeled 2 oz nalgene bottles, placed on ice, and taken to the laboratory, and frozen.

Nitrogen analysis. Approximately 5 g of the current year's foliage was weighed to the nearest 1.0 mg, dried at 60°C for 72 h, and used to determine total, alcoholic soluble, and insoluble foliar nitrogen content using standard microkjeldahl techniques. Percent water content of the foliage also was determined.

Terpene analysis. In the laboratory, needles were removed from the twigs, weighed to near 300 mg, and ground with a mortar and pestle under liquid nitrogen. The resulting fine powder was extracted with ether exhaustively (four times) in a Ten Broecke homogenizer, and filtered through glass wool. Fenchone was added as an internal standard (118 μg), and the extracts were diluted to a standard volume and injected into a capillary, gas-liquid chromatograph (GLC). GLC was performed using a 25m x 0.25mm ID fused silica column coated with OV-101. One μl was injected into a Sigma II B gas chromatograph (Perkin-Elmer) equipped with a split injector and a flame ionization detector, both of which were equilibrated at 200°C. Temperature programming was from 70°C to 190°C at 6°C/min with a 12-minute hold time at the upper temperature. The amounts of the individual components were calculated by a Hewlett-Packard recorder/integrator interfaced with the GLC amplifier. Identification was by co-chromatography of the components with authentic standards using columns of widely different polarities (OV-101 and Carbowax 20M), and by comparison of the mass spectra generated by a Finnigan GS/MS equipped with a SE-54 column with those from the National Bureau of Standards. Data are expressed as mg/g dry weight tissue.

Statistical methods. Thirty chemical and physical tree parameters were used as independent variables in multiple stepwise correlation analyses. Twenty of the variables defined foliage quality characteristics of the sample trees (Table I). For each year (1981 and 1982), two correlation analyses were performed using adult female budworm average dry weight and adult male budworm average dry weight as dependent variables. The dependent variables were determined by calculating average female and male adult budworm dry weights per sample tree based on those adults that were raised as larvae on sample trees. Additionally, sample trees were ranked with regard to their overall susceptibility to

Table I. The dependent and independent variables used in the
multiple-correlation analysis.

Independent variables Tree physiological and physical variables	Foliage quality variables
Age Height Dbh Crown diameter Crown ratio Bole radius Five-year growth increment Average internode length (for 1981 and 1982) Time of budbreak Sapwood/basal area ratio	Tricyclene[a], alpha-pinene, beta-pinene, camphene, myrcene, carene, limonene, linalool, terpinolene, bornyl acetate, citronellyl acetate, geranyl acetate, cadinene, sesquiterpenes 3 and 4, total terpenes, terpene evenness[b], percent tissue water content, total nitrogen content, ratio of soluble to insoluble nitrogen

Dependent variables
For female budworm: average adult female dry weight
For male budworm: average adult male dry weight

a
  Terpenes expressed as mg/g dry weight

b
  Based on Simpson's diversity index (6)

Table II. Female multiple correlation model and associated
ANOVA table (1981).

Source	df	SS	MS	F	R^2	P<F
Analysis of variance table						
Regression	5	272.012	54.402	6.16	0.583	0.001
Residual	22	194.415	8.837			
Total	27	466.427				

Variable entered	Coefficient[1]	F	P<F
β-pinene	−0.005	16.4	0.0005
Five-year growth increment	5.46	11.8	0.002
Terpene evenness	45.36	9.7	0.005
1981 internode length	−1.23	6.1	0.022
Terpinolene	0.034	4.5	0.045
Intercept = −15.6602			

1
  Unstandardized

budworm as determined by the average adult female budworm dry
weight associated with each tree. Trees producing a high adult
female budworm dry weight were considered more susceptible than
trees producing a low adult female budworm dry weight. Larger
females produce more offspring in the subsequent generation, thus
increasing the number of budworms and presumably the damage on the
tree in the following season (4). To determine if the relative
resistance of sample trees changed from 1981 to 1982, Spearman's
rank correlation procedure was used to compare the resistance
ranking of the sample trees between years (5). If a significant
positive correlation was found, we assumed that the relative
resistance ranking (or level) among sample trees was not significantly different between years. That is, a relatively resistant
tree in 1981 remained relatively resistant in 1982 (i.e. approximately, the same resistance rank was maintained between years).

Results

The model generated for 1981 shows that female dry weight production was inversely related to beta-pinene concentration and
internode length (Table II). However, female biomass was positively related to the most recent five-year growth increment, terpene evenness, and terpinolene. In this case, terpene evenness
is an expression of the evenness in the quantitative distribution
of terpenes in the foliage among the sampled trees. A positive
correlation between this variable and female biomass production
suggests that trees with an even distribution of terpenes in the
foliage are more susceptible than if this distribution were
skewed. Three of the five variables (60%) which entered the
model were secondary metabolites. Nitrogen did not enter into
the model as an important variable in determining female dry
weight production.
     In 1982, female dry weight production was inversely related
to bornyl acetate amd an unidentified sesquiterpene (Table III).
Neither of these were associated with reduced biomass production
in the 1981 model. Greater internode growth, evenness in the
quantitative distribution of terpenes, myrcene, camphene, and
larger crown diameters were positively associated with female
biomass production. Five of the seven variables (71%) which
entered into the model were secondary metabolites, and none of
the variables estimating nitrogen metabolism entered the model as
important determinants of female biomass production.
     For 1981, average male dry weight production was inversely
related to the concentration of limonene and carene, and to trees
with large diameter at breast height (DBH) and large sapwood/
basal area ratios (Table IV). Male biomass production was positively related to terpene evenness and high bornyl acetate and
camphene concentrations. Five of the seven variables (71%) in
the model were secondary metabolites. Nitrogen did not enter
into the model as an important determinant of male biomass production.
     For 1982, average male dry weight production was inversely
related to bornyl acetate concentration, terpene evenness, percent water content of the foliage, and tree age (Table V). Male

Table III. Female multiple correlation model and associated
ANOVA table (1982).

Source	df	SS	MS	F	R^2	P<F
Analysis of variance table						
Regression	8	550.455	68.807	4.49	0.4796	0.006
Residual	39	597.379	15.317			
Total	47	1147.834				

Variable entered	Coefficient[1]	F	P<F
1982 internode growth	2.094	11.8	0.001
Terpene evenness	39.526	6.3	0.016
Bornyl acetate	−0.005	5.8	0.021
Myrcene	0.008	5.5	0.024
Sesquiterpene 3	−0.008	4.0	0.052
Camphene	0.003	3.7	0.062
Crown diameter	1.120	3.2	0.082
Intercept = −19.5237			

[1] Unstandardized

Table IV. Male multiple correlation model and associated ANOVA
table (1981).

Source	df	SS	MS	F	R^2	P<F
Analysis of variance table						
Regression	7	35.321	5.046	5.99	0.677	0.007
Residual	20	16.869	0.084			
Total	27	52.190				

Variable entered	Coefficient[1]	F	P<F
Terpene evenness	37.74	40.1	0.0001
Limonene	−0.004	12.6	0.002
Bornyl acetate	0.001	11.2	0.003
DBH	−0.156	10.1	0.005
Camphene	0.001	5.6	0.028
Sapwood/basal area ratio	−23.65	3.1	0.09
Carene	−0.012	2.1	0.159
Intercept = −12.7186			

[1] Unstandardized

Table V. Male multiple correlation model and associated ANOVA
table (1982).

Source	df	SS	MS	F	R^2	P<F
Analysis of variance table						
Regression	10	119.406	11.941	2.73	0.445	0.014
Residual	34	148.654	4.372			
Total	44	268.060				

Variable entered	Coefficient[1]	F	P<F
1982 internode growth	1.383	15.2	0.0004
Bornyl acetate	−0.003	12.3	0.0013
Limonene	0.004	7.0	0.012
Terpene evenness	−21.969	7.0	0.013
% water content	−63.138	6.8	0.013
Tree height	0.450	4.1	0.05
Age	−0.143	3.6	0.066
Carene	0.047	2.0	0.166
Five-year growth increment	0.978	1.9	0.174
Alpha-pinene	0.001	1.4	0.237
Intercept = 88.9567			

[1]   Unstandardized

biomass production was positively associated with 1982 internode
length, limonene, carene, and alpha-pinene concentration, tree
height, and five-year growth increment. Five of the 10 variables
(50%) in the model were secondary metabolites. Nitrogen did not
enter into the model as an important determinant of male biomass
production, but interestingly, percent water content was inversely related to male dry weight production.

Finally, when the resistance rankings for the sample trees
were compared between years, a positive and significant relationship between years was found (Spearman's r=0.518, p=0.0034)
suggesting that a relatively resistant tree in 1981 remained
relatively resistant in 1982.

Discussion

It is important to note that, within a sex, the variables which
were correlated with dry weight production changed from year to
year. There was not a consistent pattern as to which variables
may be reliably used to predict budworm success (tree susceptibility). In 1981, trees with a high concentration of beta-
pinene produced smaller females, while in 1982, trees with high
concentration of beta-pinene produced smaller females, while in
1982, trees with high concentrations of bornyl acetate and an
unidentified sesquiterpene produced the smaller females. For
males, limonene and carene were inversely related to male dry
weight production in 1981, while bornyl acetate and terpene evenness were inversely related to male biomass production in 1982.
The five-year growth increment was positively correlated with
female dry weight production in 1981, but did not appear in the
1982 female model. Crown ratio was important in the 1982 female
model, but did not appear in the 1981 female model.
Additionally, the internode length (an approximation of the
amount of foliage available for the budworm) was negatively
correlated with female dry weight in 1981 and positively associated with the dependent variable in 1982. In 1982, internode
length, tree height, and the five-year growth increment were
positively associated with male dry weight in 1981 but did not
enter into the 1982 model. In 1982, internode length, tree
height, and the five-year growth increment were positively associated with male budworm weight while tree age was inversely
related to male weight. Of the last four variables only internode length entered(correlated positively with the dependent
variable) into the 1981 male model.

Within a single year, there are also differences in the
correlation models between the sexes. Such differences may be
attributable to differential foliage quality requirements of the
sexes. Sexual differences in feeding requirements and behaviors
have been reported for species of mosquitoes and tabanids (6).
Whether sexual differences in foliage requirements for
Choristoneura spp. exist presently is unknown to us.

Such results make generalizations difficult as to what truly
reduces or enhances budworm weight. However, the significant
positive relationship between the resistance rankings of the
sample trees between years is interesting. This relationship

suggests that the sample trees are maintaining their relative
position along a "resistance-susceptibility" gradient through
time (for at least two years). This seems to be the case despite
the fact that the independent (tree susceptibility) drastically
change from year to year. There may be at least two reasons why
the relative resistance-susceptibility rankings of the sample
trees remain somewhat constant while the independent variables
comprising the models predicting resistance change from year to
year. First, since there are some thirty independent variables
initially being entered into each of the correlation analyses, it
is possible that by chance alone completely different models
could be generated. Second, in order to "prevent" insect adaptation to host defenses, long-lived host trees may have been
selected to rapidly alter or vary qualitatively and quantitatively their chemical defenses to the univoltine budworm
(7,8,9,10). Therefore, it would not be unusual to find different
compounds deterring budworm in different years. Compound "A" in
one year may deter budworm feeding and may rapidly select for
those budworms which can detoxify it. Consequently, compound "A"
may rapidly become ineffective as an insect deterrent and may
even become a feeding stimulant or cue. Therefore, trees which
constantly alter, quantitatively and qualitatively, their defensive chemistry should be at a selective advantage in preventing
budworm damage. While this reason may explain why the models
from year to year are different with respect to the significant
monoterpenes, it does not adquately explain why budworm should
differentially respond to tree physical parameters nor why the
correlation models for males and females within a single year are
different. Obviously, further research into the chemical and
physical basis for budworm resistance in Douglas-fir is
warranted. Currently, investigations are continuing within our
laboratory to determine what if any role the between- and within-
tree chemical variation has on budworm success. Hopefully, these
ongoing studies will help answer some of the complex questions
raised here.

Acknowledgments

We should like to thank J. Hormer, Karen Peterson, C. Henderson,
M. Freehling and numerous undergraduate students for their help,
constant questioning, and discussion of various aspects of the
projects. This research was supported in part by NSF grant DEB
7927067 to RGC and by the Canada/United States Spruce Budworms
Program, and Accelerated Research, Development and Application
Program sponsored by the USDA Forest Service.

Literature Cited

1. Cates, R.; Redak, R.; Henderson, C. Ch. 1, In "Plant
   Resistance to Insects"; Hedin, P. A. Ed.; ACS Symposium
   Series No 208, American Chemical Society, Washington, D.C.,
   1983; p. 3.

2. Cates, R. G.; Redak, R. A.; Henderson, C. B. <u>Zeit. Entomol.</u> 1983, 96:973.
3. Redak, R. A.; Cates, R. G. <u>Oecologia</u> 1984, 62:61.
4. Redak, R. A. MS Thesis. University of New Mexico, Albuquerque, 1982.
5. Peilou, E. C. Ecological Diversity. 1975. John Wiley and Sons, New York, New York. 165 pp.
6. Chapman, R. F. The Insects: Structure and function, 3rd Edition. 1982; Harvard University Press, Cambridge, Mass. 419 pp.
7. Cates, R. G.; Redak, R. A. In "Chemical Mediation of Coevolution"; Spencer, K. Ed.; University of Chicago Press, Chicago, Il., 1985; In Press.
8. Schultz, J. C. Ch. 3, In "Plant Resistance to Insects"; Hedin, P. A. Ed.; ACS Symposium Series No. 208, American Society, Washington, D.C., 1983; p. 37.
9. Schultz, J. C.; Nothangle, P. J.; and Baldwin I. T. <u>Amer. J. Bot.</u> 1983, 69:753.
10. Schultz, J. C.; Baldwin, I. T. In "Proc. Forest Defoliator-Host Interactions: A comparison between Gypsy Moth and Spruce Budworms." U.S.D.A. Forest Service, 1983, General Technical report NE-85.

RECEIVED October 3, 1985

BIOCHEMICAL MECHANISMS OF PLANT-INSECT INTERACTIONS

10

Possible Mechanisms for Adverse Effects of L-Canavanine on Insects

Douglas L. Dahlman[1] and Milan A. Berge[2]

[1] Department of Entomology, Graduate Center for Toxicology, University of Kentucky, Lexington, KY 40546
[2] Graduate Center for Toxicology, University of Kentucky, Lexington, KY 40546

L-Canavanine is a nonprotein amino acid that can function in higher plants as a protective allelochemical.
Insects can be characterized as being canavanine-utilizers, -resistant or -sensitive. Canavanine-resistant
Heliothis virescens clear canavanine from hemolymph
faster and have LC_{50} and LD_{50}'s significantly greater
than canavanine-sensitive Manduca sexta. Canavanine-
sensitive M. sexta are unable to metabolize L-ornithine
and develop edema and ultimately die, because of canaline inhibition of ornithine-oxo-acid amino trans-
aminase. Edema also can be produced with high levels of
dietary D-ornithine. Canavanine has a marked capacity
to interfere with the reproductive potential of canavanine sensitive insects, reducing fertility and fecundity
in adult M. sexta and causing easily discernable histo-
logical lesions in the oöcytes. Dietary canavanine
also reduces fertility and fecundity in Oncopeltis
fasciatus.

L-Canavanine [2-amino-4-(guanidinooxy) butyric acid] is a structural analogue of L-arginine which replaces arginine as a substrate
in most metabolic reactions. It is one of over 240 nonprotein
amino acids identified from higher plants (1). Canavanine and its
metabolite, L-canaline exhibit intrinsic toxicity and potent anti-
metabolic properties in a variety of organisms (2-5). The growth-
inhibitory and insecticidal properties of canavanine also are well
known (3, 6-8). It seems certain that additional insights into the
actions of canavanine in insects will be forthcoming as the result
of studies on specific physiological systems as well as other
insect species.
    The most detailed investigations on the metabolism of canava-
nine and the adverse physiological and biochemical effect of canavanine in insects have been conducted in our laboratories and in
the laboratories of our associates. We have found that insects
thus far tested can be placed into one of three groups, based on
their biochemical ability or lack of ability to detoxicate and/or

0097-6156/86/0296-0118$06.00/0
© 1986 American Chemical Society

employ canavanine. There are the canavanine-utilizers exemplified by Caryedes brasiliensis [Coleoptera: Bruchidae]. Rosenthal and coworkers have described a series of experiments which demonstrate the capacity of this insect to not only detoxicate but to employ canavanine as a primary nitrogen source for amino acid synthesis (9-14). A second group of insects, typified by Heliothis virescens [Lepidoptera: Noctuidae], are canavanine-resistant and possess substantial ability to detoxicate canavanine, even though they may show certain effects, if the level of canavanine exceeds the capacity of their detoxication mechanisms. We call the third group canavanine-sensitive. Manduca sexta [Lepidoptera: Sphingidae] is the most extensively studied insect in this final group. This canavanine-sensitive species readily acylates canavanine and incorporates it into newly synthesized protein. In addition, canavanine participates in various other reactions in which arginine normally participates.

Our experience with M. sexta is quite extensive and this species has proven to be ideal for many studies requiring an organism sensitive to canavanine. More recently, we have found that H. virescens has a remarkable capacity to detoxicate canavanine. Thus, we now have available for our studies two lepidopteran species, both of which are easily and continuously reared in the laboratory on artificial diet.

Toxicology

Canavanine Toxicity. When canavanine was incorporated into a diet, a concentration as low as 3 mM (528 ppm) caused reduced growth in 5th-instar M. sexta and produced greater than 75% mortality when the 3 mM diet was offered during the entire larval period (7). The LC_{50} for the 5th-instar has been determined to be 5.2 mM (15). Manduca sexta larvae, fed approximately the LC_{50} dose, failed to pupate and even at approximately one-half of the LC_{50} dose (2.5 mM) all pupae were deformed and 80% failed to complete adult development (16). In contrast, a concentration of 200 mM canavanine was necessary to produce significant change in the growth pattern of 5th-instar H. virescens. Even at the calculated LC_{50} of 300 mM, pupae resulting from surviving larvae appeared normal even though their weight was only one-half that of the controls (17). In spite of their resistance to canavanine, H. virescens were found to be relatively sensitive to a number of other nonprotein amino acids such as L-2,4-diaminobutyric acid, L-azetidine-2-carboxylic acid, L-methionine-DL-sulfoximine, and seleno-DL-methionine (17). Thus we presume this insect does not possess a universal detoxication mechanism for nonprotein amino acids.

Clearance of Canavanine. The reported LD_{50} (parenteral injection) of canavanine for newly ecdysed 5th-instar M. sexta was 1.0 mg/g fresh body weight (15). Clearance of canavanine from M. sexta hemolymph was biphasic with a $T_{\frac{1}{2}}$ of 5 hr for the first phase and approximately 30 hr for the second phase (Figure 1). A clearance curve of this configuration helps to explain why canavanine is more toxic to M. sexta simply because the compound is present to become incorporated into protein and to antagonize various arginine-dependent reactions. It must also be remembered that chronic exposure

results in a continuous input of canavanine which replaces a portion of that metabolized and/or eliminated. In fact, the hemolymph
titer of canavanine in 5th-instar M. sexta fed a 2.5 mM canavanine-
containing diet ranged between 1 and 1.5 mM during most of the
active feeding period (18). A second study which used several
additional canavanine concentrations also found the titer of canavanine in the hemolymph to be approximately one-half the diet
concentration (15).

On the other hand, the LD_{50} for 5th-instar H. virescens was
10.7 mg/g (17), more than 10-fold greater than M. sexta. Moreover,
the clearance of canavanine showed only a single phase in H. virescens with a $T_{\frac{1}{2}}$ of 135 min (17) (Figure 1), whether the administered
dose was 1 or 5 mg/g and whether or not the larvae had been pre-
exposed to canavanine-containing diet. In addition, H. virescens
cleared canaline, a potentially poisonous metabolite of canavanine
(19), even more rapidly than the parent compound with a $T_{\frac{1}{2}}$ of 42
min (17) (Figure 1).

It was thought initially that the rapid clearance of canavanine by H. virescens could be explained by elimination of the
parent compound either via passage directly through the body without absorption or through efficient removal from the hemolymph of
the absorbed compound by the Malpighian tubules. Alternatively, it
could be rapidly metabolized by fat body or other tissues. We have
shown by means of automated amino acid analysis that only 0.34 μmol
(0.6% of the ingested dose) of canavanine was recovered from frass
collected during the final 24 hr period of a chronic, 3-day feeding
study on a diet containing 150 mM canavanine (17). We calculated
that 56 μmol of canavanine were ingested during the final 24 hr and
a total of 111 μmol were consumed during the entire 3-day feeding
period. Therefore, canavanine was not being effectively eliminated
as the parent compound. In addition, the amount of canavanine
metabolized by an in vitro incubation mixture of gut contents
failed to substantiate the hypothesis that extracellular microorganisms in the midgut lumen were responsible for the rapid removal
of canavanine from the larval body.

The hemolymph titer of canavanine in H. virescens larvae,
which at the time of hemolymph collection weighed 320 mg, was only
4.74 mM, less than 3% of the dietary concentration. The hemolymph
weight of a larva this size is approximately 28% of its fresh
weight or 90μl, if we assume the specific gravity of the hemolymph
to equal 1. This means that only 0.43 μmol or 0.38% of the total
ingested amount remained in the hemolymph after a 3 day feeding
period on a 150 mM canavanine-containing diet.

The above information can be summarized as follows: canavanine 1) is cleared rapidly from the hemolymph, 2) is not eliminated
in quantity in the frass, 3) is not detoxicated by gut microorganisms, and 4) is found only in low titer in the hemolymph of larvae
feeding on canavanine-containing diets. Therefore, we conclude
that H. virescens has an appreciable, but yet unidentified, capacity
to metabolize canavanine.

Protein Synthesis. One possible metabolic sink for canavanine is
incorporation into de novo-synthesized protein. Within 24 hr M.
sexta incorporated approximately 3.5% of an injected dose of L-
[guanidinooxy-^{14}C]canavanine into protein precipitated with tri-

chloroacetic acid (9). Our unpublished data have shown that the
rate of incorporation reached a maximum by that time. We also know
that the major portion of the labeled protein was associated with
the integument and cuticular materials that constitute the body
wall tissues. In contrast, the canavanine-resistant H. virescens
incorporated canavanine into protein at only 5% the rate observed
in M. sexta with the maximum rate of incorporation occurring within
4 hr post-injection. Thus, it would appear that incorporation into
protein is not a major reservoir for canavanine in H. virescens.

Effect of Canaline, a Canavanine Metabolite

Canaline is the product of the hydrolytic cleavage of canavanine
with the simultaneous formation of urea. Canaline is an ornithine
analogue which also shows neurotoxicity in the adult M. sexta where
it adversely affects central nervous system functions (19). It also
is a potent inhibitor of vitamin B_6-containing enzymes (20-22). It
forms a stable Schiff base with the pyridoxal phosphate moiety of
the enzyme and drastically curtails enzymatic activity. Pyridoxal
phosphate-containing enzymes are vital to insects because they
function in many essential transamination and decarboxylation reactions. Ornithine is an important metabolic precursor for insect
production of glutamic acid and proline (23).

Elevation of Ornithine Titer. Administration of canavanine by
either the parenteral or oral route resulted in an abnormally high
concentration of ornithine in the hemolymph of M. sexta. This was
first reported by Racioppi and coworkers (18) (Figure 2) and later
confirmed by Lenz (15), who also showed a similar increased ornithine titer in M. sexta fed canaline-containing diets. Elevated
titers of ornithine also have been observed in the hemolymph of H.
virescens reared on 150 mM canavanine-containing diet (24). Either
parenteral injection or dietary administration of canavanine to M.
sexta larvae inhibited markedly the pyridoxal phosphate-dependent
enzyme, ornithine-oxo-acid amino transferase [EC 2.6.1.13], which
transfers the amino group on ornithine to 2-oxyglutaric acid (25).
The unstable product, L-glutamate-semialdehyde, converts spontan-
eously to either L-glutamate or L- '-pyrroline-5-carboxylate,
depending upon the enzyme system involved. Inhibition of ornithine-
oxo-acid amino transferase is the most probable cause for the
accumulation of ornithine in the hemolymph of both M. sexta and H.
virescens.

Edema. Manduca sexta larvae fed a diet which contained 2.5 mM
canavanine and 25 mM ornithine, arginine or citrulline, developed
edema (26). Their hemolymph volume nearly doubled (27) and individual weights increased by as much as 60%. In a study on the
potentiation of canavanine induced developmental anomolies in M.
sexta by 29 different amino acids (26), it was shown that those
compounds which accentuated the biological activity of canavanine
have or can be readily converted to have the following: an   -
carboxyl and -amino group; a carbon skeleton of no less than 2 nor
more than 4 carbon atoms, and an -amino group. The edematose
condition only occurred in the presence of canavanine and excessive
amounts of those compounds having 3 or 4 carbon atoms in the carbon
skeleton and the other above mentioned functional groups.

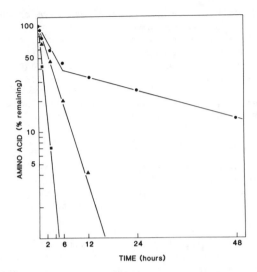

Figure 1. Clearance of L-canavanine and L-canaline from insect hemolymph. Clearance was determined by automated amino acid analysis involving a single sample for each time point using the pooled hemolymph of five larvae. The administered dose to H. virescens was 5 mg L-canavanine/g fresh body weight ( ) or 3.8 mg L-canaline/g fresh body weight ( ). Manduca sexta received 1 mg L-canavanine/g fresh body weight ( ). Redrawn from (17).

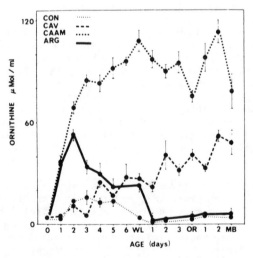

Figure 2. Concentrations of hemolymph ornithine in M. sexta larvae fed throughout the 5th-instar on diet containing either control (CON), 25 mM arginine (ARG), 2.5 mM canavanine (CAV) or 2.5 mM canavanine + 25 mM arginine (CAAM). Each data point represents the statistical mean from at least 5 individuals and the vertical bars represent +SE and are not shown where the SE is less than the height of the symbol. "Reproduced with permission from Ref. 18. Copyright 1981, Pergamon Press."

Interestingly, the osmotic concentration of the hemolymph of edematose larvae was nearly normal (28). Apparently M. sexta larvae have no satisfactory alternative means to eliminate the accumulated ornithine and edema results either from an active attempt to maintain osmotic homeostasis in the hemolymph or as the result of passive water retention.

We determined D-ornithine to yield only 15% as much product as an equal concentration of L-ornithine when incubated with M. sexta ornithine-oxo-acid amino transferase. We reasoned therefore that it might be possible to mimic the effect of elevated L-ornithine titer associated with canavanine toxicity by feeding a diet supplemented with D-ornithine. Indeed, edematose larvae were obtained from larvae reared on a 50 mM D-ornithine-containing diet (Table I) whereas the same concentration of L-ornithine had no deleterious effect. Greater mortality prior to pupation was observed in the larvae reared on the 50 mM D-ornithine diet than on a diet containing 2.5 mM canavanine plus 25 mM arginine (CAAM). A diet containing only 25 mM D-ornithine produced a lesser degree of edema along with decreased larval mortality. It is important to note that 50 mM D-lysine, the higher homologue of D-ornithine, did not produce edema even though it caused 80% mortality in the pupal stage (Table I). This circumstantial evidence suggests that at least some of the adverse effects of canavanine result from secondary biochemical imbalance brought about by metabolites of the parent compound.

Table I. Effect of D- and L-Amino Acids on Manduca sexta Development[1]

Treatment	Maximum Weight (g)	Mortality (%) During			Adult Emergence (%)
		Larval Stage	(ILPE)[2]	Pupal Stage	
Control Diet	10.0 + 0.2	0	0	13	87
CAAM[3]	12.5 + 0.4	0	76	20	4
50 mM D-Ornithine	12.3 + 0.4	73	27	--	0
50 mM L-Ornithine	9.8 + 0.8	7	0	7	87
50 mM D-Lysine	10.2 + 0.2	7	13	80	0
50 mM L-Lysine	10.0 + 0.3	0	0	13	87

[1] Diet preparation and rearing procedures were as described in (26).
[2] ILPE = Incomplete Larval-Pupal Ecdysis.
[3] CAAM = 2.5 mM L-canavanine + 25 mM L-arginine.

Effects of Canavanine on Reproductive Potential

Even though the effects of canavanine and/or its metabolic products
on macromolecular synthesis and enzyme inhibition have been the
subject of most of our investigations, descriptive studies on the
effects of canavanine suggest that examination of other areas may
become equally fruitful. ,Some of the most sensitive effects resulting from chronic exposure of insects to canavanine have been
observed in the development of reproductive tissues, the formation
of oöcytes, and the resulting fecundity and fertility of the eggs.
Manduca sexta. In the early studies on the effects of canavanine
on M. sexta, it was noted that the ovarial mass (ovaries, oöcytes
and associated fat body and tracheal system) was greatly reduced
(33% of the control) in female moths obtained from larvae which had
fed upon canavanine-containing diet during their final larval stadium. Other metabolic allies of canavanine had much less effect
(16).
     Palumbo has examined the effects of chronic exposure to cana-
vanine in the last larval stadium on the reproductive capacity of
adult M. sexta. He observed that diet concentrations as low as 0.5
mM (89 ppm) resulted in reduced egg fertility and fecundity (29).
It was noted that fewer oöcytes were produced and the egg chorions
either failed to form or were of such structure that the eggs "dim-
pled" after being laid. Embryonic development either did not begin
or failed at a relatively early point (30). Examination of thin
sections of the oöcytes showed obvious gross structural differences
between tissues from control and canavanine treated insects (Figures 3 and 4). The trophocytes in control tissue possessed relatively uniform stained cytoplasm with few vesicles whereas the
cytoplasm of the trophocytes from canavanine-treated insects contained numerous vesicles of various sizes. Likewise, the contents
of control oöcytes were relatively uniform and contained vesicles
ranging up to 6 microns in diameter. In contrast, oöcytes from
canavanine-treated insects contained many more vesicles of much
larger size (20 microns or more in diameter). Many of the vesicles
in the treated oöcytes appeared to contain some type of material
which stained only lightly with toludine blue. We speculate that
such oöcytes are not viable.
     Preliminary studies to determine sperm viability and effect of
canavanine on production of accessory gland products were inconclusive. However, experiments in which canavanine-treated males were
mated with control females indicated some reduction in fertility
whereas canavanine-treated females mated with normal males showed
reduction in both fertility and fecundity (29). This observation
would be expected if the effect of canvanine is expressed both
directly in the gamate cell and in the trophocytes which provide
nourishment for the developing oöcyte. It seems unlikely that
sufficient canavanine remained in the adult to have a direct effect
at the time of oöcyte formation. Rather, it is supposed that canavanine exerted its influence during the early stages of development
of the reproductive structures in the final larval stadium. The
mass of tissue at this time is less than 2 mg as compared to 400 mg
in the fully developed adult (31).

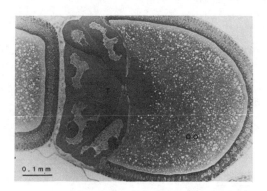

Figure 3.   Follicle at stage five of development from control of
M. sexta newly emerged adult illustrating trophocytes (T)
emptying cytoplasm into the oöcyte (OO).

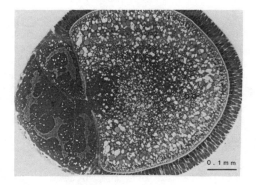

Figure 4.   Follicle at stage five of development from canavanine-
treated M. sexta newly emerged adult illustrating trophocytes
(T) emptying cytoplasm into the oöcyte (OO). Note the large
number of vesicles in the cytoplasm and the increased number and
size of vesicles in the oöcyte as compared to the control in
Figure 3.

Dysdercus koenigii. The antigonadal effects of canavanine on
Dysdercus koenigii [Hemiptera: Pyrrhocoridae] have been examined
(32). The amino acid was offered in the water to newly emerged
adults. During an exposure time of 6 days, a solution of 40 ppm
produced significant reduction in fecundity and complete loss of
fertility. Even 20 ppm reduced egg hatch to 20% compared to nearly
100% for the control (Figure 5). As little as 80 ppm caused significant adult mortality within 9 days. In a second experiment newly
emerged adults were treated for either 2 or 3 days and then mated
with untreated adults of the opposite sex. Some reduction in
fertility was observed when the adults received as little as 40 ppm
canavanine in their water supply for only 2 days. The effect was
time and concentration dependent but was independent of sex in that
fertility was reduced equally whether only the male or only the
female received canavanine.

Periplaneta americana. Koul (33) also studied the effects of canavanine on fertility of the American cockroach, Periplaneta americana
[Orthoptera: Blattidae] by offering dog biscuits soaked in various
aqueous concentrations of canavanine to the adult cockroaches.
During the 60 day test, no significant effects were observed on the
body weight and food consumption index, yet roaches feeding on
biscuits which had been soaked in a 0.05% solution of canavanine
produced fewer oötheca and reduced numbers eggs/oöethca as compared
to roaches fed untreated biscuits. The time of hatch was extended
2 weeks and the percent hatch was only 17% that of the control.
Even a 0.02% concentration yielded a hatch of only 47% compared to
that of the control. Koul also reported that the oöthecal capsules
were deformed and easily damaged.

    In both of the above studies (32,33) the insects were treated
as adults, thus demonstrating the possible direct effect of canavanine on the reproductive tissues. Indeed, it was observed that the
ovaries of the canavanine-treated insects were very small and
undifferentiated. In contrast to M. sexta, these insects are long
lived and undergo several reproductive cycles. Thus, there is
greater opportunity for canavanine to exercise adverse effects on
these tissues.

Oncopeltus fasciatus. Recent work in our laboratory with the large
milkweed bug, Oncopeltus fasciatus [Hemiptera: Lygaeidae] has shown
results similar to the work of Koul (32). The insects were reared
on unsalted sunflower seeds and the canavanine was offered in the
water supply. We were only able to rear 8% of the newly hatched O.
fasciatus nymphs to the adult stage on a 62.5 ppm concentration of
canavanine (Table II). However, a concentration of 31 ppm gave
nearly as many adults as the control, even though development time
was longer. Insects from the 31 ppm treatment laid fewer eggs on
the 1st day of oviposition and the eggs had a lower percent hatch,
thereby reducing the mean potential offspring to approximately one-
half that of untreated insects (Table II).

Table II. Effect of Continuous Feeding of L-Canavanine on Growth
and Reproduction of Oncopeltis fasciatus[1]

Conc. (ppm)	Adults (%)	Mean Days To Adults	No. Eggs Laid on 1st day	Egg Hatch (%)	Mean Potential Offspring
125.0	0	---	---	---	---
62.5	8	41.5 + 1.4	0	---	---
31.2	86	33.0 + 0.3	7 + 2	74	4.9
0	92	29.3 + 0.2	13 + 3	83	10.7

[1]Groups of 10 newly hatched nymphs were placed in one-half pint
paper cans and fed unsalted sunflower seeds and water containing
various concentrations of L-canavanine. Five groups of nymphs were
started for each concentration of canavanine. This study was
terminated shortly after nymphs had reached adulthood.

Table III. Effect of L-Canavanine on Reproductive Capacity of
Oncopeltis fasciatus Administered Continuously During Entire
Adult Life

Conc. (ppm)	n	Mean Days to Death	Mean Total Eggs/	Mean Eggs/ /Day	Egg Hatch (%)	Mean Potential Offspring
1000	4	26.2 + 2.6	83 + 46	5.1 + 1.9	12.6	10 + 6
500	3	30.3 + 0.7	275 + 62	13.7 + 2.4	43.1	110 + 27
250	4	40.5 + 4.2	395 + 41	13.7 + 1.2	41.5	164 + 17
125	4	43.8 + 9.0	506 + 177	16.1 + 1.2	66.7	338 + 118
62.5	4	42.4 + 8.1	618 + 137	20.0 + 2.0	84.0	519 + 115
31.2	4	56.8 + 8.2	1102 + 262	22.8 + 2.4	76.7	845 + 201
0	2	63.0 + 8.0	902 + 122	17.4 + 0.2	72.6	655 + 89

[1]Newly emerged adults were placed in one-half pint paper cans, one
female and two males, supplied with sunflower seeds and water containing various concentrations of canavanine. Five replications
were started but fewer were used for data because of early death or
poor egg laying by some females.

We also used a protocol where treatment was initiated with
newly emerged adult milkweed bugs. The results of this experiment
are shown in Table III. The preoviposition period (10-12 days) was
not affected by canavanine but all other parameters measured showed
concentration-dependent effects. Thus canavanine increased mortality, reduced the total number of eggs laid (fecundity) as well as
the percent hatch (fertility). Once oviposition began, O. fasciatus
normally deposited eggs daily until death. The preoviposition
period was subtracted from the total number of live days before the

rate of egg deposition of (eggs/ /day) was determined. The calculated mean potential offspring [the product of percent hatch (fertility) and total eggs laid (fecundity)] provides a sensitive test
for evaluation of chemical effects on reproductive potential. His-
tological studies have not yet been conducted and we presently have
no information on the metabolism of canavanine in O. fasciatus.

It should be clear from this work with four species from three
different insect orders that canavanine has potential as an antigonadal substance. Information on the specific mode of action is
wanting but the physical evidence presented suggests that such
studies should be conducted.

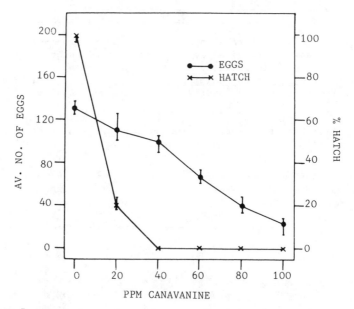

Figure 5. Mean fecundity and fertility of D. koenigii offered
various concentrations of L-canavanine during their adult life
up to the time of laying the 1st batch of eggs. Bars: range of
values obtained. "Reproduced with permission from Ref. 32.
Copyright 1983, Dr. W. Junk Publishers."

Acknowledgments

The authors gratefully acknowledge the support of the U.S. Department of Agriculture Agreements 59-2213-1-1-763-0 and 84-CRCR-1-
1469, and National Institutes of Health Grant AM 17322. This paper
is published as No. 85-7-108 of the Kentucky Agricultural Experiment Station, Lexington 40546.

Literature Cited

1.   Rosenthal, G. A. "Plant Nonprotein Amino and Imino Acids.
      Biological, Biochemical, and Toxicological Properties";
      Academic Press: New York, 1982; pp. 2, 95-107.

2.   Fowden, L.; Lea, P. J.; Bell, E. A.   Adv. Enzymol.   1979, 50, 117-175.
3.   Rosenthal, G. A.; Bell, E. A.   In: "Herbivores: Their Inter-action with Secondary Plant Metabolites"; Rosenthal, G. A.; Janzen, D. H., Eds.; Academic Press: New York, 1979; pp. 353-385.
4.   Bell, E. A. Prog. Phytochem.   1980, 7, 171-196.
5.   Fowden, L. Perspect. Exptl. Biol.   1976, 2, 263-272.
6.   Rosenthal, G. A. Quart. Rev. Biol.   1977, 52, 155-178.
7.   Dahlman, D. L.; Rosenthal, G. A. Comp. Biochem. Physiol.   1975, 51A, 33-36.
8.   Rosenthal, G. A.   In "Plant Resistance to Insects"; Hedin, P. A., Ed.; American Chemical Society: Washington, D.C., 1983; pp. 279-290.
9.   Rosenthal, G. A.; Dahlman, D. L.; Janzen, D. H.   Science 1976, 192, 256-258.
10. Rosenthal, G. A.; Janzen, D. H.; Dahlman, D. L.   Science 1977, 196, 658-660.
11. Rosenthal, G. A.; Dahlman, D. L.; Janzen, D. H.   Science 1978, 202, 528-529.
12. Rosenthal, G. A.; Hughes, C. G.; Janzen, D. H.   Science 1982, 217, 353-355.
13. Rosenthal, G. A. J. Chem. Ecol.   1983, 9, 803-815.
14. Rosenthal, G. A.; Janzen, D. H. J. Chem. Ecol.   1983, 9, 1353-1361.
15. Lenz, C. J. M.S. Thesis, University of Kentucky, Lexington, 1983.
16. Rosenthal, G. A.; Dahlman, D. L. Comp. Biochem. Physiol.   1975, 52A, 105-108.
17. Berge, M. A.; Rosenthal, G. A.; Dahlman, D. L. Pestic. Biochem. Physiol. (in press).
18. Racioppi, J. V.; Dahlman, D. L.; Neukranz, R. K. Comp. Bio-chem. Physiol.   1981, 70B, 639-642.
19. Kammer, A. E.; Dahlman, D. L.; Rosenthal, G. A. J. Exp. Biol.   1978, 75, 123-132.
20. Rosenthal, G. A. Eur. J. Biochem.   1981, 114: 301-304.
21. Rahiala, E.-L.; Kekomäki, M.; Jänne, J.; Raina, A.; Räihä, N. C. R. Biochem. Biophys. Acta.   1971, 227, 337-343.
22. Rahiala, E.-L. Acta Chem. Scand.   1973, 27, 3861-3867.
23. Cochran, D. G.   In "Insect Biochemistry and Function"; Candy, D. J.; Kilby, B. A., Eds.; Chapman and Hall: London, 1975; pp. 177-281.
24. Berge, M. A. unpublished observations.
25. Racioppi, J. V.; Dahlman, D. L. unpublished observations.
26. Dahlman, D. L.; Rosenthal, G. A. J. Insect Physiol.   1982, 28, 829-833.
27. Dahlman, D. L. Ent. Exp. Appl.   1978, 24, 327-335.
28. Racioppi, J. V.; Dahlman, D. L. Comp. Biochem. Physiol.   1980, 67C, 35-39.
29. Palumbo, R. E.; Dahlman, D. L. J. Econ. Entomol.   1978, 71, 674-676.
30. Palumbo, R. E. M.S. Thesis, University of Kentucky, Lexing-ton, 1976.
31. Sroka, P.; Gilbert, L. I. J. Insect Physiol.   1971, 17, 2409-2419.
32. Koul, O. Ent. Exp. Appl.   1983, 34, 297-300.
33. Koul, O. Z. ang. Ent.   1983, 96, 530-532.

RECEIVED October 3, 1985

11

Consequences of Modifying Biochemically Mediated Insect Resistance in *Lycopersicon* Species

George G. Kennedy

Department of Entomology, North Carolina State University, Raleigh, North Carolina 27695-7630

Phenols, alkaloids and methyl ketones have been demonstrated to play important roles in the mediation of insect resistance in plants of the genus Lycopersicon. Although they are considered likely candidates for genetic manipulation, through plant breeding to achieve insect resistance, their use is complex. The actual level of resistance which results from the presence of a given level of a particular compound often depends upon the larger chemical context within which that compound occurs in the plant (i.e. the presence of other biologically active compounds). In addition the introduction of biochemically-based resistance to a particular insect species may have unanticipated and undesirable effects on other, nontarget insect species. Examples of these phenomena are presented and discussed in relation to their implications for utilizing insect resistant tomato cultivars for crop protection.

The cultivated tomato, Lycopersicon esculentum Mill., is attacked by a number of very serious arthropod pests which, at present, are controlled largely through the use of insecticides (1,2). In recent years, research has been directed towards the identification and mechanistic understanding of arthropod resistant tomato germplasm for use in developing tomato cultivars resistant to arthropods. As a result, several potentially useful sources of arthropod resistant germplasm have been identified (3 and references therein) and an understanding of the complexity of biochemical factors mediating arthropod resistance in members of the genus Lycopersicon is beginning to emerge. Several chemicals including the catecholic phenols rutin and chlorogenic acid (4,7), the glycoalkaloid α-tomatine, (8–10) and the methyl-ketones 2-tridecanone and 2-undecanone (11–17) which occur in tomato foliage have been

0097–6156/86/0296–0130$06.00/0

implicated in arthropod resistance. This paper will describe the
known effects of these compounds on selected insect pests of tomato
and describe some of complexities involved in manipulating the
presence and amounts of these chemicals in the tomato plant to
enhance the level of insect resistance. Emphasis will be on three
insect species for which the greatest amount of information is
available: the tobacco hornworm, Manduca sexta L.; the tomato
fruitworm, Heliothis zea (Boddie), and the Colorado potato beetle
Leptinotarsa decemlineata (Say).

The Catecholic Phenols Rutin and Chlorogenic Acid

Foliar phenolics of tomato have been investigated extensively
as possible factors in the growth inhibitory effects of tomato
foliage on H. zea larvae. The catecholic phenols chlorogenic acid
and rutin account for over 60 percent of the total phenolic content
of tomato foliage (6, 7) and occur in both the leaf lamella and the
tips of the type VI glandular trichomes (Figure 1) which abound on
tomato foliage. Duffey and Isman (5) reported that rutin was the
major phenolic (80-90%) in the tips of type VI glandular trichomes
of L. esculentum, with chlorogenic acid and other caffeic acid
conjugates comprising the remainder of the catecholic derivatives.
Isman and Duffey (6) found that catecholic phenols in tomato foliage
can act in an additive fashion to inhibit growth of H. zea larvae
and that a serial dilution of a total phenolic extract of tomato
foliage gave rise to a dose-response (growth inhibition) in larvae
similar to that produced by pure chlorogenic acid or rutin.
These results and those of Elliger et al. (4), which are based
on the incorporation of phenolics in artificial diet, lend support
to the hypothesis that tomato phenolics contribute to the antibiotic
potential of tomato foliage to H. zea. However, in a subsequent
study in which they examined phenolic content of foliage from
different tomato cultivars and larval growth on those cultivars,
Isman and Duffey (7) found no correlation between the phenolic
content and larval growth, despite the fact that the levels of
phenolics in the foliage were comparable to those causing
significant reductions in larval growth on diet.   The level of
growth on foliage (even that containing low levels of phenolics) in
those experiments was low compared to growth on artificial diet,
suggesting that other factors either interfered with or over-rode
the biological activity of the phenolics.
The interaction that occurs between phenolics and dietary
protein and affects phenolic toxicity may explain the lack of
correlation between phenolic content and larval growth on tomato
foliage. Duffey (18) has found that the growth inhibitory effects
of rutin on H. zea larvae are dependent upon the amount of protein
in the diet. The growth inhibitory effects of a given concentration
of rutin in diet increased with the protein content of the diet over
the range of protein levels tested (0.6-4.8%). Further, different
types of protein varied in their ability to enhance rutin-induced
toxicity. If this type of interaction operates in foliage as it
does in artificial diet, it may make the breeding of tomato
cultivars with elevated levels of H. zea resistance by selecting for
high levels of foliar phenolics difficult to achieve because

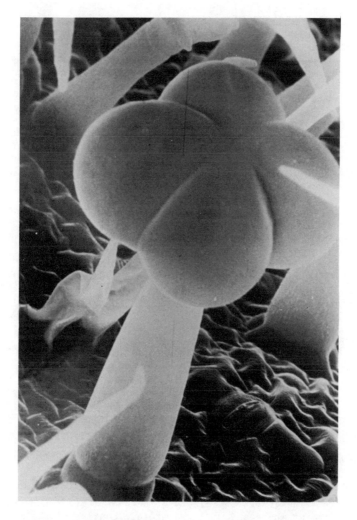

Figure 1. Type VI glandular trichomes of the cultivated
tomato L. esculentum.

selection would have to be for high levels of foliar protein as
well. Confounding the problem of using high phenolic and high leaf
protein levels as a basis for H. zea resistance in tomato is the
fact that the levels of both phenolics and proteins in tomato
foliage are highly variable both within and between plants of the
same and different cultivars (e.g. protein levels range from 0.04%
to 6.0% fr wt.; 18).

α-Tomatine

The glycoalkaloid α-tomatine has been implicated in the resistance
of tomato to both H. zea (4, 10, 18) and to L. decemlineata (8).
α-Tomatine has been found in all Lycopersicon species surveyed (19,
20). It is present in both tomato foliage and fruit, although the
concentration of α-tomatine in the fruit decreases as the fruit
matures (4, 10, 20). For at least two tomato species which have
been examined (L. esculentum and L. hirsutum f. glabratum), foliar
α-tomatine has been found associated with the leaf lamellae but not
the tips of the type VI trichomes (5, 21).

Heliothis zea.   α-Tomatine has been shown to inhibit growth and
development of H. zea when added to artificial diet at
concentrations comparable to those found in tomato foliage (4, 18).
Elliger et al. (4), for example, reported that foliage of the
commercial tomato cultivars `Ace´ and `Campbell 29´ contained 0.76
and 0.61 mg α-tomatine per gm fresh weight, respectively, while
foliage of the highly insect resistant accession PI134417 of the
wild tomato species Lycopersicon hirsutum f. glabratum contained
2.45 mg α-tomatine per gm fresh weight. All of these values were in
excess of the amount of α-tomatine required to reduce larval growth
of H. zea to a level 50% of that of the controls in artificial diet
studies (ED_{50}=0.40 mg α-tomatine/gm fr. wt. diet). However, Elliger
et al. (4) made no attempt to relate larval growth on foliage with
the α-tomatine content of the foliage.
     In studies by Juvik and Stevens (10), performance of H. zea
larvae on tomato fruits of varying developmental ages from three
accessions of Lycopersicon was related to the α-tomatine content of
the fruits. The $\log_n$ of α-tomatine content was negatively
correlated with larval growth rate and adult weight but positively
correlated with time to pupation and mortality (10), suggesting
α-tomatine contributes to the resistance of immature tomato fruits
to H. zea larvae. The growth of H. zea larvae on tomato foliage,
however, could not be related to the α-tomatine content of the
foliage of different cultivars (refs. 4 & 5 cited in 7). A partial
explanation for this may lie in the ability of equimolar quantities
of phytosterols to alleviate the toxicity of α-tomatine to H. zea
(18), especially since Campbell and Duffey (9) found the amounts of
free sterol or sterol ester and of α-tomatine in the foliage of
different tomato cultivars to vary independently. Thus, to the
extent that α-tomatine is involved in the antibiosis of tomato
foliage towards H. zea larvae, current evidence indicates that the
interaction between α-tomatine and phytosterols is at least as
important as the absolute quantity of α-tomatine present in
determining the level of antibiosis.

Determining the potential value of exploiting elevated levels
of α-tomatine as a basis for enhanced resistance to H. zea is
further complicated by the finding that elevated levels of α-
tomatine in the diet of H. zea adversely affect Hyposoter exiguae,
an ichneumonid parasitoid attacking H. zea (9, 22). However, the
toxic effects of α-tomatine on H. exiguae are also alleviated by the
presence of equimolar or supramolar quantities of certain
phytosterols in the diet of its host, H. zea (9). Thus, the
available data indicate that in order to utilize α-tomatine as a
basis for foliar resistance to H. zea, one would have to breed for
plants having high levels of α-tomatine and low levels of certain
phytosterols. If this were achieved, the toxic effects of α-
tomatine might extend beyond the H. zea population and adversely
affect the parasitoid H. exiguae, a potentially important natural
enemy of H. zea (23, 24).

Leptinotarsa decemlineata. α-Tomatine also acts as a feeding
deterrent for L. decemlineata (25) and the tomatine content of the
wild tomato species that are resistant to L. decemlineata is
generally higher than that of commercial tomato cultivars (8).
Nonetheless, the evidence that α-tomatine is responsible for the
resistance of any tomato genotypes to L. decemlineata is
inconclusive, but suggestive.
    In experiments in which adult beetles were fed L. esculentum
leaf disks infused with α-tomatine, Sinden et al. (8) found a
positive relationship between inhibition of feeding and α-tomatine
content of the infiltrated leaf disk. Further, they demonstrated
that feeding by adult beetles on foliage of L. esculentum and L.
hirsutum f. glabratum (PI134417) plants of different ages or from
plants grown under different daylength regimes was negatively
correlated with α-tomatine content. The correlation was higher for
L. esculentum foliage (r=-0.897) than for L. hirsutum f. glabratum
foliage (r=-0.613). However, because their studies revealed
significant differences between the two plant species, they
concluded that although important, α-tomatine content was less
important in determining feeding rate on L. hirsutum f. glabratum
than on L. esculentum. Subsequent research (12, 13, 26, 27) has
implicated the methyl ketone 2-tridecanone as an additional factor
in the resistance of L. hirsutum f. glabratum to L. decemlineata.
Unlike the situation with H. zea, where α-tomatine exerts an
antibiotic effect (i.e. chronic toxicity) when ingested by the
larvae, α-tomatine appears to operate against L. decemlineata adults
primarily as a feeding deterrent. None of the available data
indicate any interaction between α-tomatine and phytosterols or any
other compounds sufficient to obscure the feeding deterrent effects
of α-tomatine on potato beetle. However, experiments specifically
designed to detect such interactions apparently have not been
conducted.
    If the role of α-tomatine in resistance to L. decemlineata is
verified, it should be relatively easy to select for high levels of
foliar α-tomatine in a tomato breeding program because the genetic
variation in α-tomatine is controlled by the segregation of 2 co-
dominant alleles at a single locus (28). One possible limitation on
the utility of α-tomatine-mediated resistance to L. decemlineata is

that the α-tomatine content of the foliage is not only influenced by daylength, but also increases as the plant ages (8). Since L. decemlineata is often a serious pest of young plants, to be most useful in combating the potato beetle, high levels of α-tomatine would have to be present in the foliage of young plants.

The Methyl-Ketones 2-Tridecanone and 2-Undecanone

The wild tomato L. hirsutum f. glabratum PI134417 is highly resistant to M. sexta, L. decemlineata and H. zea because its foliage is toxic to the larvae of all three species (16, 26, 29). The type VI glandular trichomes (30) present on the foliage play an important role in the acute toxicity of the foliage to these insects, although their importance in the resistance to M. sexta and L. decemlineata is greater than that to H. zea (Table I).

Table I.   Effect of Removing Type VI Glandular Trichome Tips on Toxicity of L. hirsutum f. glabratum (PI134417) Foliage.

| | Percent Mortality[1] | | | |
| | H. zea | | M. | L. |
	Test I[2]	Test II[3]	sexta	decemlineata
L. hirsutum f. glabratum				
Trichomes Intact	83a	65a	100a	97a
Tips Removed	34b	47b	10b	15b
L. esculentum	18b	16c	7b	17b

[1] Mortality recorded at 96 hr for H. zea and 72 hr for M. sexta and L. decemlineata. Mean separation vertical P<0.05.
[2] Dimock and Kennedy 1983.
[3] Farrar and Kennedy, unpublished.

The contents of the type VI trichome tips are acutely toxic to these three species, as well as several others by virtue of the presence of rather large amounts of 2-tridecanone (3, 11, 12, 13, 16, 26). In tests in which neonate larvae were confined on 2-tridecanone-treated filter paper, the LC_{50} values for M. sexta, H. zea and L. decemlineata were 17.05, 17.1 and 26.9 ug/cm^2 treated surface, respectively. Virtually all (99%) of the 2-tridecanone present occurs in association with the glandular trichomes (16). On average, there are 6.3 ng 2-tridecanone per trichome tip or 44.6 ug 2-tridecanone/cm^2 of leaflet surface for PI134417. This is enough to account for the observed acute toxicity of the foliage to neonate larvae of the three species. In contrast, foliage of L. esculentum contains a mean of only 0.1 ug 2-tridecanone/cm^2 of leaflet (12, 16).

Manduca sexta and Leptinotarsa decemlineata. For 2-tridecanone to be effective as a resistance mechanism, larvae on the resistant plant must be exposed to lethal quantities of the toxin. The role

of 2-tridecanone as a principal factor in the lethality of PI134417
foliage to neonate M. sexta and L. decemlineata larvae was confirmed
by experiments showing a close association between the abundance of
2-tridecanone and the resistance of foliage from PI134417 plants of
different ages and/or grown under different light regimes, as well
as in experiments examining resistance and 2-tridecanone levels in
plants from segregating F_2 and backcross populations from crosses
between L. esculentum and PI134417 (12-14, 31-33). The experiments
with L. decemlineata also revealed the presence of resistance
factors, other than 2-tridecanone, which are associated with the
lamellae of PI134417 foliage. This lamella-based resistance is
manifest in PI134417 only in foliage from which the glandular
trichomes have been removed and in F_1 (L. esculentum x PI134417)
progeny which have levels of 2-tridecanone (0.5 ug/cm^2) that are
comparable to those of L. esculentum. It is expressed as a gradual
accumulation of mortality throughout larval development (Table II,
13). When high levels of 2-tridecanone are present, the young
beetle larvae are killed by the 2-tridecanone, often without having
fed on the foliage, and the presence of the lamella-based resistance
is obscured. The specific factors responsible for the lamella-based
resistance are not known, although it is quite possible that α-
tomatine is involved (8).

Table II.   Survival of L. decemlineata Reared on
            PI134417, L. esculentum and Their F_1
            Progeny.

Plant Population	Percent Survival To	
	4 Days	Adult
L. esculentum	85a	43a
PI134417	0b	0b
F_1 (L. esculentum x PI134417)	79a	11c

Mean separation vertical P<0.01.

Heliothis zea. As with L. decemlineata, the overall resistance of
PI134417 to H. zea involves factors associated with the glandular
trichomes as well as those associated with the leaf lamallae (Table
I). It is not known whether the same factors associated with the
leaf lamallae condition resistance to H. zea and L. decemlineata.
It is clear, however, that the glandular trichome-mediated portion
of the resistance to H. zea involves much more than simply high
levels of 2-tridecanone.
     Although 2-tridecanone is acutely toxic to neonate H. zea by
contact and fumigant action (16) and the quantities associated with
the foliage of PI134417 are potentially lethal, most neonate H. zea
larvae do not experience a lethal dose of 2-tridecanone. Rather,
neonates experience a narcotizing dosage resulting from their
exposure to 2-tridecanone vapors which are released by the
trichomes; these larvae become paralyzed, but most (ca. 80%)
subsequently recover and begin feeding on the foliage (16). Since
initial knockdown by 2-tridecanone kills ca. 20% of the larvae, 2-

tridecanone accounts for at least a part of the trichome-mediated
resistance to H. zea. Exposure of neonate larvae to low
concentrations of 2-tridecanone on filter paper produced the same
knockdown/recovery response as observed on PI134417 foliage,
indicating that 2-tridecanone was responsible for the knockdown
effect (34). Further, larvae which have been exposed either in
vitro or on PI134417 foliage to narcotizing dosages of 2-tridecanone
and recovered are more tolerant to 2-tridecanone in subsequent
exposures and more tolerant to the commercial insecticide carbaryl
than comparable larvae which had no prior exposure to 2-tridecanone
(Tables III & IV, 34).

Table III.   Effect of 2-Tridecanone Pretreatment on
             Tolerance of H. zea Larvae to 2-Tridecanone
             in Subsequent Exposures.

| 2-Tridecanone Treatment | 2-Tridecanone Pretreatment | |
| | $(N\ Moles/cm^2)$ | |
$N\ Moles/cm^2$	0	8.5
0	0.0	0.0
17.0	31.5	0.0 *
25.5	50.5	31.5 *
34.0	78.5	42.0 *
42.5	77.0	68.5 NS

* Values= % Mortality. Difference between pretreatment rates significant $P{<}0.05$. Reprinted with permission
from Ref. 34. Copyright 1984 Dr. W. Junk Publishers.

Table IV.   Effect of Pretreatment With 2-Tridecanone
            on Toxicity of Carbaryl to H. zea.

Pretreat (16 hrs)	Treatment[1]	Mean Percent Mortality[2]
H_2O	Carbaryl	67a
L. esculentum foliage	Carbaryl	63a
Diet	Carbaryl	49a
2-Tridecanone	Carbaryl	17b
PI134417 Foliage	Carbaryl	18b
Diet	Diet	1c

[1] 25 ug technical carbaryl (1 napthyl methyl-
   carbamate) topically applied to 3rd instar larvae.
   (See Kennedy 1984 for details).
[2] Mean separation $P{<}0.01$; Duncan's multiple range
   test. Reprinted with permission from Ref. 34. Copyright
   1984 Dr. W. Junk Publishers.

Although not definitive, these findings suggest that 2-
tridecanone is an inducer of mixed-function oxidase activity (MFO).

Subsequent experiments have confirmed this hypothesis, showing a 2-fold increase in cytochrome P_{450} levels in H. zea larvae confined for 48 h on PI134417 foliage or on diet containing 2-tridecanone as compared to artificial diet lacking 2-tridecanone or L. esculentum foliage (35).

The induction of MFO's by 2-tridecanone suggests that the inappropriate use of tomato cultivars selected for high levels of 2-tridecanone-mediated resistance to M. sexta and L. decemlineata without concurrent selection for high levels of resistance to H. zea, attributable to other plant factors, could result in H. zea being more difficult to control with certain insecticides.

In addition to 2-tridecanone, the type VI trichome tips of PI134417 also contain several other ketones, including 2-undecanone (36). 2-Undecanone is less abundant in the type VI tips than 2-tridecanone (1.1 ng/tip vs 6.3 ng/tip, respectively (21)) and is less toxic to H. zea in contact toxicity tests (LC_{50}=64.2 vs 17.1 ug/cm^2 treated surface, respectively (15)).

Since most H. zea larvae on PI134417 foliage die only after having consumed substantial quantities of foliage, the effects of 2-undecanone and 2-tridecanone on H. zea larvae were further investigated by feeding larvae artificial diet containing either or both 2-undecanone and 2-tridecanone (37). Consistent with previous studies (16, 24), larvae reared from egg hatch to pupation on diet containing 0.3% 2-tridecanone (wt/wt=plant rate) suffered a small but significant increase in mortality during the first 48 hr (15% on 2-tridecanone vs 3% on control P<0.0001). Surviving larvae developed normally to pupation and adult emergence. Doubling the concentration of 2-tridecanone in the diet increased mortality during the first 48 hr to 50%. 2-Undecanone at levels comparable to those found in PI134417 foliage (0.055%) is also toxic to H. zea when ingested, but unlike 2-tridecanone, its toxicity is manifested as the production of deformed pupae and high levels of mortality during the pupal stage (72% vs 2% in control). Since the toxic effects of 2-undecanone were manifest even when 2-tridecanone was present in the diet, the induction of cytochrome P_{450} which results from exposure of H. zea larvae to sublethal dosages of 2-tridecanone has no detoxicative effect on 2-undecanone.

Present information suggests that a high level of H. zea resistance in tomato could be obtained if cultivars could be selected having very high levels of 2-tridecanone (ca. 3 times those found in PI134417) or high levels of 2-undecanone (1-2 times those in PI134417). Assuming there are no interactions between these compounds and other chemicals in the plants which would attenuate the toxicity of these ketones, plants with high levels of 2-tridecanone would cause extensive mortality of neonate larvae while those with high levels of 2-undecanone would cause extensive pupal mortality. In areas where damaging populations of H. zea invade the crop from sources outside the crop, 2-tridecanone-mediated resistance would be more valuable because it would prevent the development of large populations of late instar larvae. However, since H. zea already has the ability to detoxify moderate dosages of 2-tridecanone, selection for populations of H. zea able to tolerate 2-tridecanone mediated resistance could be rapid under some circumstances. Since the effects of 2-undecanone-mediated

resistance would be manifest only in the number of moths produced in
the tomato crop, this resistance would be most valuable in
situations where the insect population cycles through more than one
generation in the tomato crop before reaching damaging levels.

Conclusions

A number of biologically-active chemicals which have potential for
genetic manipulation in the breeding of insect resistant tomato
cultivars occur in plants of the genus **Lycopersicon.** However,
because of interactions among plant constituents involving some of
these chemicals, the level of resistance to a particular insect
which results from the presence of a given level of a particular
compound cannot be predicted without additional information on the
abundance of interacting constituents and an understanding of the
nature of the interactions themselves. Similarly, where the
resistance-mediating chemicals are concentrated in the glandular
trichome tips (e.g. 2-tridecanone and 2-undecanone), the level of
resistance manifest by the plant is determined by the interaction
between glandular trichome density (i.e. No. trichome/cm^2) and the
amount of the chemicals per trichome tip. Thus, breeding for insect
resistance attributable to a particular biochemical mechanism is
likely to require selection for appropriate levels of each of an
array of interacting chemical and plant characters.
    A further complicating factor is that the introduction of a
particular biochemically-mediated resistance to one insect species
may have unanticipated and undesirable effects on other nontarget
insect pests and beneficial species. Given these complexities, a
thorough understanding of the interactions leading to the expression
of useful levels of insect resistance is an important prerequisite
to the development of tomato cultivars having high levels of
multiple insect resistance.

Acknowledgments

I thank Sean Duffey, Department of Entomology, U. C. Davis, for
generously sharing his unpublished data and for numerous valuable
and thought provoking discussions. Portions of this work were
supported by the U. S. Department of Agriculture under Agreement No.
58-7B30-1-281 with N. C. State University and U.S.D.A. Competitive
Research Grant No. 83-CRCR-1-1241 to P. Gregory and G. G. Kennedy.
Paper No. 9959 of the Journal Series of the North Carolina
Agricultural Research Service, Raleigh, NC 27695-7601. The use of
trade names in this publication does not imply endorsement by the
North Carolina Agricultural Research Service of the products named,
or criticism of similar ones not mentioned.

Literature Cited

1. Lange, W. H.; Bronson L. **Ann. Rev. Entomol.** 1981, 26, 345–
      71.
2. Kennedy, G. G.; Romanow, L. R.; Jenkins, S. F.; Sanders, D. C.
      J. Econ. Entomol. 1983. 76, 168–73.

3. Kennedy, G. G.; Yamamoto, R. T. Entomol. Exp. Appl. 1979, 26, 121-6.
4. Elliger, C. A.; Wong, Y.; Chan, B. G.; Waiss, A. C., Jr. J. Chem. Ecol. 1981, 7, 753-8.
5. Duffey, S. S.; Isman, M. B. Experientia 1981, 37, 574-6.
6. Isman, M. B.; Duffey, S. S. Entomol. Exp. Appl. 1982, 31, 370-6.
7. Isman, M. B.; Duffey, S. S. J. Amer. Soc. Hort. Sci. 1982, 107, 167-70.
8. Sinden, S. L.; Schalk, J. M.; Stoner, A. K. J. Amer. Soc. Hort. Sci. 1978, 103, 596-600.
9. Campbell, B. C.; Duffey, S. S. J. Chem. Ecol. 1981, 7, 927-46.
10. Juvik, J. A.; Stevens, M. A. J. Amer. Soc. Hort. Sci. 1982, 107, 1065-9.
11. Williams, W. G.; Kennedy, G. G.; Yamamoto, R. T.; Thacker, J. D.; Bordner, J. Science 1980, 207, 888-9.
12. Kennedy, G. G.; Dimock, M. B. In IUPAC Pesticide Chemistry, Human Welfare and the Environment; Miyamoto, J., Ed.; Pergamon Press; New York, p. 123-8.
13. Kennedy, G. G.; Sorenson, C. E.; Fery, R. L. Proc. Symp. Ecology and Management of the Colorado Potato Beetle. Mass. Agric. Expt. Stn. Bull. 1985 (in press).
14. Fery, R. L.; Kennedy, G. G.; Sorenson, C. E. Hort. Science 1984, 19, 86.
15. Dimock, M. B.; Kennedy, G. G.; Williams, W. G. J. Chem. Ecol. 1982, 8, 837-42.
16. Dimock, M. B.; Kennedy, G. G. Entomol. Exp. Appl. 1983, 263-8.
17. Farrar, R.; Kennedy, G. G., unpublished data.
18. Duffey, S. S., unpublished data.
19. Roddick, J. G. Phytochemistry. 1974, 13, 9-25.
20. Juvik, J. A.; Stevens, M. A.; Rick, C. M. HortScience. 1982, 17, 764-6.
21. Ave, D.; Gregory, P., unpublished data.
22. Campbell, B. C.; Duffey, S. S. Science. 1979, 205, 700-2.
23. Michelbacher, A. E.; Essg, E. O. Calif. Agr. Exp. Stn. Bull. 1930, 625.
24. Puttler, B. Ann. Entomol. Soc. Amer. 1961, 54, 25-30.
25. Sturchow, B.; Low, I. Entomol. Exp. Appl. 1961, 4, 133-42.
26. Kennedy, G. G.; Sorenson, C. E. J. Econ. Entomol. 1985, 78: 547-551.
27. Sorenson, C. E. M.S. Thesis. North Carolina State University, Raleigh, NC 27695, 1984.
28. Juvik, J. A.; Stevens, M. A. J. Amer. Soc. Hort. Sci. 1982, 107, 1061-5.
29. Kennedy, G. G.; Henderson, W. R. J. Amer. Soc. Hort. Sci. 1978, 103, 334-6.
30. Luckwill, L. C. Aberdeen Univ. Studies. 1943, No. 120.
31. Kennedy, G. G.; Yamamoto, R. T.; Dimock, M. B.; Williams, W. G.; Bordner, J. J. Chem. Ecol. 1981, 7, 707-16.
32. Fery, R. L.; Kennedy, G. G. HortScience 1983, 18, 169.
33. Schwartz, R. F.; Snyder, J. C. HortScience 1983, 18, 170.
34. Kennedy, G. G. Entomol. Exp. Appl. 1984, 35, 305-11.

35. Riskallah, M. R.; Farrar, R.; Kennedy, G. G.; Hodgson, E.
    unpublished data.
36. Clement, P.; Bordner, J. unpublished data.
37. Farrar, R.; Kennedy, G. G.; Ave´, D.; Gregory, P. unpublished
    data.

RECEIVED August 9, 1985

12

Biochemical Bases of Insect Resistance in Rice Varieties

R. C. Saxena[1]

International Centre of Insect Physiology and Ecology, P.O. Box 30772, Nairobi, Kenya

Rice plant allelochemics greatly influenced behavior
and physiology of striped stemborer (SSB), Chilo
suppressalis (Walker), brown planthopper (BPH),
Nilaparvata lugens (Stål), and green leafhopper (GLH),
Nephotettix virescens (Distant). Steam distillate
extract of resistant 'TKM6' rice plants inhibited
SSB oviposition, hatching, and larval development,
but susceptible 'Rexoro' plant extract induced
oviposition. Oviposition inhibitor in 'TKM6' was
identified and synthesized. Extracts of susceptible
varieties attracted BPH while those of resistant
varieties repelled. Less individuals settled and
fed on susceptible 'TN1' plants sprayed with extracts
of resistant 'ARC 6650' and 'Ptb 33' varieties than
on plants sprayed with 'TN1' extract. Similarly,
application of extract of resistant variety 'ASD7'
to 'TN1' plants disrupted GLH feeding. Topical
application of extracts of resistant rice varieties
and nonhost barnyard grass killed more BPH adults
than extracts of susceptible plants. BPH biotypes
differed in their relative vulnerabilities to
extracts of resistant varieties. Gas chromatographic
analysis of resistant and susceptible plant extracts
revealed qualitative and quantitative differences in
volatiles. Major amino acids differed quantitatively
in 'Mudgo', 'ASD7', and 'TN1' varieties and their
ability to stimulate BPH feeding. All test varieties
were suitable for BPH egg-laying but hatchability was
low on resistant varieties and barnyard grass. Trans-
aconitic acid in barnyard grass parenchymatous tissue
reduced BPH hatchability, but not that of GLH.

Susceptibility or resistance of plants is the result of a series of
interactions between plants and insects which influences the ultimate

[1]Current address: IRRI, P.O. Box 933, Manila, Philippines.

degree of establishment of insect populations on plants (1, 2, 3).
The factors which determine insect establishment on plants can be
categorized into two groups: (1)   Insect responses to plants, and
(2) Plant characters influencing insect responses. The insect
responses include orientation, feeding, metabolic utilization of
ingested food, growth of larvae to adult stage, adult longevity, egg
production, oviposition and hatching of eggs. Unfavorable biophysical or biochemical plant characters may interrupt one or more of
these insect responses, inhibiting establishment of an insect population on a plant and rendering it resistant to infestation and injury.

Generally, morphological or biophysical resistance factors in
plants interfere with insects' vision, orientation, locomotion,
feeding, mating, or oviposition mechanism. Biochemical factors are
far more important in imparting resistance to insects as plant
chemicals affect insect behavior and physiology in a number of ways.
These factors may be nutritionally-based or may include nonnutritio-
nal chemicals, called "allelochemics" (4), that affect insect
behavior, growth, health, or physiology. Some of these allelochemics
have been found to be associated with repellence, feeding deterrence,
toxicity, or other adverse effects on insects (5). Allelochemics
with such negative effects on the receiving organism (insect) are
termed "allomones" (6).   Other allelochemics may serve as attractants, feeding stimulants and sometimes may interact with nutrients
to increase their metabolic utilization by insect (7, 8). Allelochemics which give adaptive advantage to the receiving organism are
called "kairomones" (6). Plant resistance may not only be affected by
the presence of allelochemics, but even their absence in host plants
may alter drastically their resistance to one insect pest species or
the other. Potential nutritiveness of a plant may also influence its
susceptibility to pests.

A sound basis for developing resistant varieties should be
oriented towards identifying the resistance imparting chemicals and
using them as cues in breeding programs. However, determination of
these factors is often intricate because adequate bioassays are
necessary before chemical differences can be identified as the cause
of resistance. This is often difficult because of a lack of information on the behavioral and physiological responses of the pests to
the plant biochemicals. Also, continual evolution of biotypes of
certain insect pest species capable of overcoming specific host plant
resistance underscores the need for a better understanding of the
basis of host plant resistance or susceptibility.

The cultivated rice, Oryza sativa L., is attacked by more than
100 insect species, of which about 20 are major pests (9). Together
they infest all parts of the plant at all growth stages, and a few
transmit virus diseases (10). In addition, a number of insects
attack stored rice.

Varietal resistance in rice has provided a highly practical
approach to controlling insect problems. Concerted efforts for
utilizing insect resistance in rice improvement program began about
two decades ago at the International Rice Research Institute (IRRI).
Since then, several thousand rice varieties and accessions have been
screened for resistance to insect pests and the selected resistant
varieties utilized in breeding improved rices. However, systematic
research in establishing bases of resistance to insect pests started

only within the last decade. The available information pertains
solely to stem borers, the brown planthopper (BPH), Nilaparvata
lugens (Stål), and the green leafhopper (GLH), Nephotettix virescens
(Distant), which are major pests of rice crop in tropical Asia.

Stem Borers. Early attempts linked the silica content of rice plants
with resistance to stem borers (11, 12), and the varietal differences
in the total silica content were recorded in a few cases (13, 14).
Larvae feeding on high silica-containing rice varieties, e.g.,
'Yabami Montakhab', were reported to exhibit typical antibiotic
effects and had worn out mandibles (15). However, wearing out of
mandibles due to a high silica content is not likely to sufficiently
impair the feeding activity of larvae on a rice plant because at
every successful molt, the larvae are endowed with new sets of
mandibles.
     Munakata and Okamoto (16) identified "oryzanone" (p-methyl
acetophenone) as a larval attractant, and benzoic and silicic acids
as larval growth inhibitors in the rice plant. A higher nitrogen
content and a greater percentage of starch were recorded in certain
varieties susceptible to the yellow stem borer, Scirpophaga
incertulas (Walker), than in resistant ones (14). Incorporation of
fresh plant water-extract of susceptible and resistant varieties into
artificial diets showed that the striped stem borer (SSB), Chilo
suppressalis (Walker), larvae had a definite preference for the
susceptible 'Rexoro' variety over the resistant 'TKM6' and suffered
poor growth on 'Taitung 16', another resistant variety (17). Antibiosis in 'TKM6' and 'Taitung 16' was attributed to biophysical and
biochemical factors, respectively. However, whether these factors
influenced other responses of the insect during the process of its
establishment on the host was not investigated.
     C. suppressalis moths distinctly prefer for oviposition the
susceptible variety 'Rexoro' over the resistant 'TKM6'. A general
association between several morphological and anatomical characters
such as length of internodes, length and width of flag leaf, culm
height, diameter, etc., in the rice varieties and stem borers was
recorded (15). Although each of these characters appeared to
contribute to borer resistance, none by itself appeared to be the
main cause of resistance. For example, tall varieties because of
their height might be more attractive to egg-laying moths. Similarly, a hairy leaf blade might act as physical deterrent to oviposition.
However, removal of hairs from the leaf of the resistant 'TKM6'
variety did not make it more attractive for oviposition by SSB moths
(15).
     Biochemical basis of susceptibility or resistance of rice
varieties to C. suppressalis has been studied in detail at IRRI
recently (18). Allelochemics, mainly plant volatiles, were obtained
from the ground leaf sheath tissue by steam distillation. The
distillate was extracted with diethyl ether and after vacuum evapora-
tion, a yellow oily residue was recovered which had the characteristic odor of each respective variety. The oily extracts of 'Rexoro'
(susceptible) and 'TKM6' (resistant) varieties were tested as such,
or as acid, basic, and neutral fractions for eliciting SSB moth's
orientational and ovipositional responses. The whole extracts were
also assayed for their effects on SSB eggs, larvae, and pupae.

When moths were allowed a choice of an untreated paper strip, a strip treated with the extract of 'Rexoro' and another strip treated with the extract of the 'TKM6' variety, they oviposited heavily on 'Rexoro' extract-treated strips, none on the 'TKM6' extract-treated strips, and very few on untreated strips. However, almost an identical number of SSB moths arrived on 'Rexoro' and 'TKM6' extract-treated strips, few moths arrived on the blank strips (Figure 1).

Comparison of SSB moth's orientational and ovipositional responses to each of the acid, basic, and neutral fractions showed that the acid fraction of 'Rexoro' extract was both a strong attractant as well as stimulated strong egg-laying response. Basic and neutral fractions of 'Rexoro' extract were ineffective. On the other hand, acid, basic, and neutral fractions of 'TKM6' extract were attractive to SSB moth, but only acid fraction stimulated some oviposition. Both basic and neutral fractions inhibited oviposition as 2 to 7 times less eggs were laid on treated paper strips than on untreated strips.

A complete reversal in the ovipositional behavior of SSB moths was obtained by spraying the 'TKM6' extract on 'Rexoro' plants on which females laid only one-tenth the number of eggs laid on normal 'Rexoro' plants (Figure 2). On the other hand, 'TKM6' plants sprayed with 'Rexoro' extract received nearly 4 times more the number of eggs laid on untreated 'TKM6' plants.

Topical application of 'TKM6' extract on eggs adversely affected embryonic development and reduced the egg hatch to ca 5% (19). In control, 98% of eggs hatched successfully. Similarly, treatment of 4th- or early 5th-instar SSB larvae with 'TKM6' extract disrupted their growth and development and all died in 3 days. Different grades of larval-pupal intermediates resulted when prepupal stage stem borer larvae were treated with 'TKM6' extract. Larvae treated with 'Rexoro' extract or with ether alone pupated normally and moths emerged successfully from them.

The above results point to qualitative and quantitative differences between the allelochemics of SSB-susceptible and resistant varieties. The insect's greater preference for egglaying on 'Rexoro' can be attributed to the presence of attractant factor(s) and oviposition inducer(s) in the acid fraction and absence of repellents and oviposition inhibitors in the basic and neutral fractions. In contrast, whatever little oviposition inducer(s) the 'TKM6' acid fraction has, its effect is masked by the inhibitors in the neutral and basic fractions. It also explains the total reversal in ovipositional behavior of SSB moths on 'TKM6' plants treated with 'Rexoro' extracts and vice versa. It may become possible to use these chemicals for interrupting SSB moth's oviposition on a rice crop, or alternatively make a trap crop so attractive that SSB moths lay all eggss on it.

Normal development of eggs and hatching, and growth of SSB larvae and pupae treated with 'Rexoro' extracts indicates that allelochemics in 'Rexoro' plants serve as SSB growth regulators. Probably, the "switch on" response of the newly-emerged, mated moths to start egg-laying on a 'Rexoro' extract-treated surface is due to an "imprinting" of message from the behavior and metabolism regulators to which the insect is exposed during younger stages in life. The

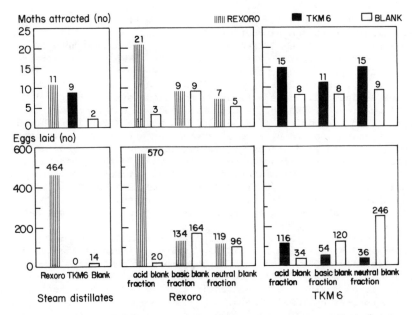

FIG. 1. Orientational and ovipositional responses of striped
stem borer, <u>Chilo</u> <u>suppressalis</u>, moths on paper strips treated
with ether extract of steam distillates of resistant (TKM6)
and susceptible (Rexoro) rice varieties.

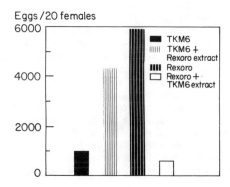

Figure 2. Ovipositional response of striped stem borer,
<u>C. suppressalis</u>, moths on plants treated with ether extract of
steam distillates of resistant ('TKM6') and susceptible ('Rexoro')
rice varieties.

insect uses this information to its advantage for recognition and colonization of the susceptible host plant.

Resistance of 'TKM6' plants to SSB is probably due to the production of certain allomones which inhibit oviposition and disrupt the insect's growth and development. Some of these factors, such as 'compound A', have recently been identified and synthesized in collaboration with the Tropical Development Research Institute (formerly Tropical Products Institute), London (20). The allomones in the resistant 'TKM6' variety thus account for inhibition of SSB oviposition and for adverse effects on egg (embryonic), larval and pupal (post-embryonic) stages. On the other hand, nine wild rices (four O. minuta J.S. Presl., four O. officinalis Wall ex. Watt, and one O. punctata Kotschy ex. Steud.) had no compound A in their plant volatiles, but a few were still toxic to SSB (21). This confirms that allomones other than compound A may also be involved in SSB resistance. The role of nutritional factors in growth of SSB larvae remains yet to be studied.

Brown Planthopper. Studies on BPH resistance in rice began at IRRI in 1966. Since then, nearly 50,000 rices have been screened and valuable sources of resistance to the pest identified. The insect thrives on high yielding susceptible varieties, but fails to feed, grow, survive, and reproduce adequately on resistant varieties (22). Nonetheless, resistance could not be traced to any morphological or anatomical peculiarity in the rice plant. Reduced feeding on resistant varieties was, therefore, attributed to either the lack of phagostimulants or to the presence of antifeedants.

Sogawa and Pathak (23) believed that resistance of 'Mudgo' rice plants was possibly due to a somewhat low concentration of amino acids, particularly asparagine, which stimulates BPH feeding. On the other hand, Kim et al. (24) indicated that the trans-aconitic acid acted as a BPH antifeedant in the nonhost barnyard grass, Echinochloa crus-galli L., a common weed in rice fields. Yoshihara et al. (25, 26) reported that soluble silicic acid and oxalic acid in the rice plant acted as BPH sucking inhibitors. But soon, it became apparent that silicic acid was a general sucking inhibitor occurring in both susceptible and resistant varieties. Likewise, oxalic acid was found to occur in both resistant and susceptible varieties, although its concentration was slightly higher in some resistant varieties. Although both these acids are water soluble, their occurrence in the phloem sap has not yet been demonstrated. Silica, which is the elemental form of the silicic acid and occurs in the soil, is more likely to be transported through the xylem vessels. On the other hand, oxalic acid, which is a product of plant's cellular metabolism, is highly toxic even to the plant tissue and, therefore, less likely to occur in the phloem. Plants rich in oxalic acid, e.g., Oxalis, sequester this metabolite as crystals in the tonoplasm of vacuoles of the cells and not as free inclusions in the cell sap (27)

Shigematsu et al. (28) identified asparagine as a sucking stimulator and β-sitosterol as a sucking inhibitor of BPH. However, none of these studies demonstrated that antifeedants extracted from the whole plants occurred principally in the phloem.

Attempts to tap the phloem in the rice plant by various techniques, including the use of laser beam (29), have not been successful

because the vascular bundles in the rice plant are scattered, and so
precise chemical analyses and bioassays of the phloem sap have not
been feasible. Even if the presence of antifeedants in the phloem
sap of resistant varieties could be demonstrated, it would not relate
to other, more vital, aspects of BPH behavior and physiology.

Saxena and Pathak (3) made systematic studies of BPH-rice plant
interactions, particularly behavioral and physiological responses
involved in BPH establishment on rice plants. They found that
resistant plants were as suitable as susceptible plants in eliciting
some responses. The interaction of all the responses determined the
overall susceptibility or resistance to the pest. Saxena and Pathak
(3) Saxena and Puma (30), and Saxena and Okech (31) also determined
the biochemical basis of suitability of rice varieties to BPH. They
found that allelochemics and nutritive balance of rice varieties were
important in eliciting optimal or suboptimal responses, thereby
affecting BPH ability to establish on rice plants. The steam distillate extracts of resistant varieties and of the barnyard grass were
repellent and, when applied topically, caused high mortalities even
at low doses. In contrast, extracts of susceptible varieties
possessed moderate to high attractance and were relatively nontoxic
to the insect. Recently, Obata et al. (32) isolated and identified
constituents of BPH attractant in the Japanese rice cultivar
'Nihonbare'.

BPH settling response on tillers of the susceptible 'TN1'
variety sprayed with the steam distillate extracts of resistant
varieties showed the same pattern of response as when actual
resistant plants were used. This indicated that treatment of the
susceptible variety with steam distillate extracts of resistant
varieties conferred resistance at least temporarily. The low amount
of honeydew excreted by BPH females on the tillers of 'TN1' plants
treated with the extract of 'ARC 6650' resistant or 'Ptb 33' variety
confirmed that the insect was unable to settle down for sustained
feeding (Figure 3). Nymphs caged on similarly treated 'TN1' plants
were unable to settle on them and suffered high mortality. Thus,
restlessness of BPH nymphs and adults on resistant plants could be
attributed to exposure to the plant volatiles which have a repellent
or toxic effect on the insect.

BPH Biotype 1, Biotype 2, and Biotype 3 maintained on 'TN1',
'Mudgo', and 'ASD7' rice varieties, respectively, showed distinct
differences in relative vulnerability to the steam distillate
extracts of their respective resistant varieties and the barnyard
grass (Figure 4) (33). The average yield of plant volatiles in some
highly resistant varieties, such as 'Ptb 33' and the barnyard grass,
was much higher than in other resistant varieties and all susceptible
varieties.

Gas chromatography of steam distillate extracts of different
rice varieties recorded 34 distinct peaks (Figure 5) (31). Many
peaks were common to all varieties but showed variation in absorbance.
However, there were also some unique peaks for certain varieties.

The exact identity of allelochemics in BPH-resistant varieties
is not yet known. However, a large group of low molecular weight
compounds, such as essential oils, particularly terpenoids, alcohols,
aldehydes, fatty acids, esters, waxes, etc., would be obtained by
steam distillation (34, 35). Obata et al. (32) have identified a

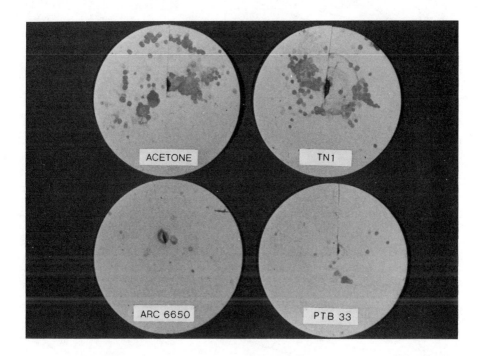

Figure 3. Filter paper disks on which honeydew of N. lugens (Biotype 1) females was collected when they fed on susceptible 'TN1' plants sprayed with steam distillate extract of resistant 'ARC 6650', 'Ptb 33', or susceptible 'TN1' rice varieties. Control plants were sprayed with acetone. Dark spots on ninhydrin-treated filter paper disks indicate the amount of honeydew excreted by females on treated rice plants.
"Reproduced with permission from Ref. 33. Copyright 1985, Plenum Publishing Corporation."

Mortality (%)

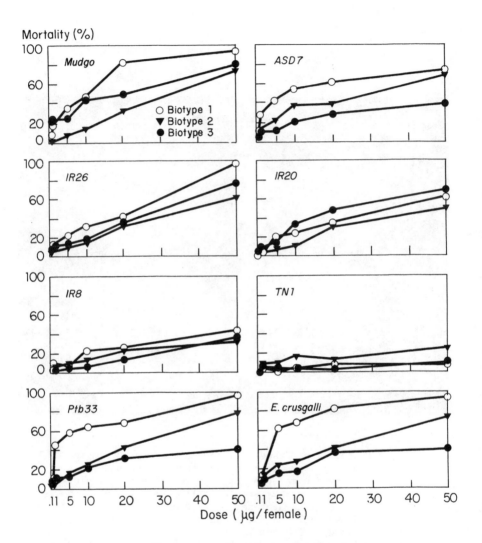

Dose (μg/female)

Figure 4. Mortality of brachypterous females of N. lugens biotypes 24 h after topical application of steam distillate extracts of different rice varieties and barnyard grass.

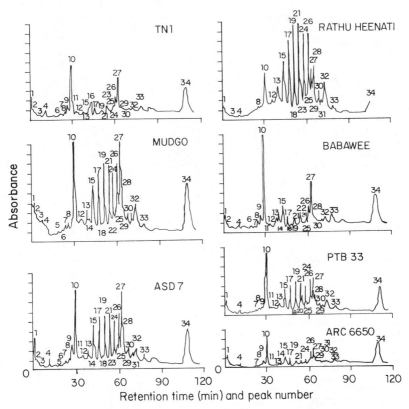

Figure 5. Gas chromatograms of volatiles of leaf sheaths of rice varieties susceptible ('TN1') and resistant to N. lugens (Biotype 1). "Reproduced with permission from Ref. 33. Copyright 1985, Plenum Publishing Corporation."

mixture of 14 esters, 7 carbonyl compounds, 5 alcohols, and 1 isocy-
nurate in the volatile attractant fraction of a BPH-susceptible
Japanese rice cultivar 'Nihonbare'.

While no major differences in total sugars and starch in BPH-
susceptible and resistant varieties have been reported, a comparison
of the free amino acids occurring in the leaf sheath tissues of 'TN1',
'Mudgo', and 'ASD7' varieties showed some quantitative differences.
Bioassays of seven of the major amino acids indicated differences in
their relative degree of phagostimulation to the three BPH biotypes
(Figure 6) (33). Thus, certain nutrients may affect suitability of
the rice plant for BPH feeding.

Significantly less BPH eggs hatch on resistant rice varieties
and on the nonhost barnyard grass than on susceptible rice varieties
(3). This could be attributed to the occurrence of injurious,
water-soluble allelochemics, such as some organic acids, which
comprise the chemical environment of BPH eggs, which are laid within
the plant tissue. Hatchability of BPH eggs was markedly impaired
when they were exposed to low concentrations (0.01-1%) of solutions
of trans-aconitic acid (Figure 7) (30). Trans-aconitic acid is present in 0.2 to 0.5% concentration in the barnyard grass (24).

In contrast, GLH eggs tolerated trans-aconitic acid levels which
were lethal to BPH eggs. In nature, GLH thrives on barnyard grass.

Green Leafhopper. Of 48,000 rice varieties and accessions from all
over the world that have been screened, about 2,000 have showed
potential for resistance to GLH (36). It damages resistant varieties
less because the leafhopper's intake of phloem sap from such
varieties is low and its feeding is restricted to xylem (37).
Recently, Khan and Saxena (38) confirmed, by using a lignin-specific
dye that is selectively translocated in xylem vessels, that GLH is
primarily a phloem feeder on susceptible varieties, but on resistant
varieties, it ingests mainly from xylem tissue. With an electronic
device, Khan and Saxena (39) also demonstrated distinct differences
in waveforms for probing, salivation, phloem feeding, and xylem
feeding on GLH-resistant and susceptible rice plants.

The shift from phloem feeding to xylem feeding on resistant
rices is not well understood. Auclair et al.(37) speculated that
these differences were due to either the presence of a feeding
deterrent in tissues adjacent to or within the sieve elements of
resistant plants or the formation of callose or slime plugs in phloem
in response to GLH probing.

Khan and Saxena (40) monitored the leafhopper's feeding behavior
on susceptible 'TN1' plants sprayed with the steam distillate extract
of resistant 'ASD7' plants with the electronic monitoring device and
the lignin-specific dye. Application of 'ASD7' extract to susceptible 'TN1' plants disrupted the normal feeding behavior of the insect
(Figure 8). Phloem feeding by the insect was significantly less on
'TN1' plants sprayed with 'ASD7' extract at 500, 1000, 2000, and
4000 ppm than on control 'TN1' plants sprayed with acetone/water
mixture (Table I). The reduced phloem feeding on the extract-treated
plants was associated with a significant increase in probing fre-
quency and an increase in duration of salivation and xylem feeding.

These results stressed the role of plant volatiles in determining insect behavior. Most plant volatiles are sparingly soluble in

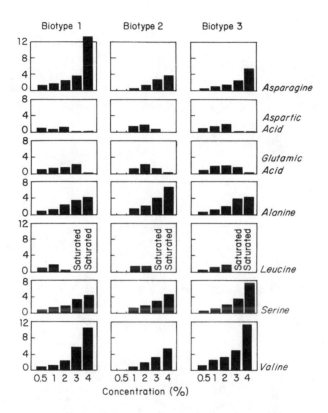

Figure 6.   Relative intake of seven amino acid solutions of different concentrations by three biotypes of N. lugens.

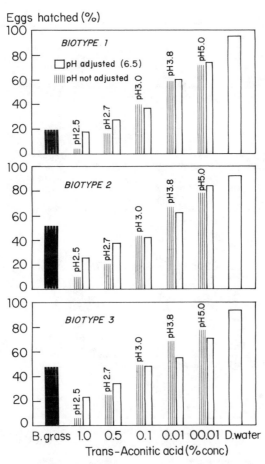

Figure 7. Effect of trans-aconitic acid, a barnyard grass chemical, on hatching of eggs of three biotypes of N. lugens.

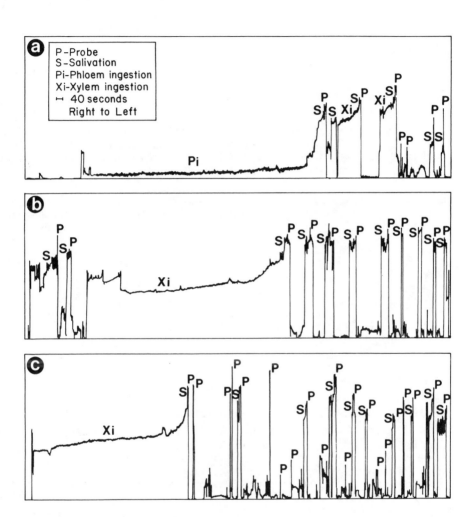

FIG. 8. Electronically recorded waveforms during N. virescens
feeding on (a) susceptible 'TN1' and (b) resistant 'ASD7' rice
seedling, both sprayed with acetone/water mixture (4:1), and
(c) 'TN1' seedlings sprayed with 4,000 ppm steam distillate
extract of 'ASD7' plants (Khan and Saxena 1985a).

Table I. Area of bluish amino acid spots indicating phloem feeding,
and number of red honeydew droplets indicating xylem feeding by
N. virescens on 'ASD7' steam distillate extract-treated 'TN1'
or control rice seedlings[a] (Khan and Saxena 1985a).

Extract concn[b] (ppm)		Area of bluish acid spots (mm^2)	Red honeydew droplets (no.)
500		805b	40c
1,000		481c	47c
2,000		390d	60b
4,000		431cd	82a
0[c] (susceptible check, 'TN1' seedlings)	Control	2326a	17d
0[c] (resistant check, 'ASD7' seedlings)	Control	203e	59b

In a column, means followed by the same letter are not significantly
different ($P<0.05$; Duncan's [1951] multiple range test).

[a]Average of 4 replications. In each replication, honeydew of 10
newly-emerged females was collected on a 9-cm-diameter filter paper
placed around the base of the seedlings. Red honeydew spots indicated xylem feeding on safranine-dyed seedlings and bluish amino acid
spots indicated phloem feeding when filter paper disks were treated
with a 0.1% ninhydrin/acetone solution.

[b]All concentrations prepared in a mixture of acetone/water (4:1).

[c]Sprayed with a mixture of acetone and water (4:1).

water and are not likely to be translocated in the vascular bundles (41). However, their odoriferous and volatile nature makes them exert a strong influence on the total chemical environment of the rice plant and hence is of ecological significance in determining the susceptibility or resistance of rice plants to insect pests.

Whittaker (4) reported that to be effective, the defense chemicals must be relatively toxic. Further, to make them available to herbivores, the plant must be, to some degree, leaky and open to some loss of materials through its surfaces. In the same manner, the rice plant volatiles may be able to interact with and affect insect responses. Volatile compounds from resistant host plants may penetrate the insect body through the cuticle or spiracles during feeding and respiration. Cuticular penetration of plant volatiles is well documented in insects. Leafhopper and planthopper nymphs, and stem borer larvae, particularly young instars, having a vestiture of poorly chitinized cuticle and relatively larger surface area because of smaller size, would therefore be vulnerable to volatile compounds of resistant rice plants, while attempting to feed on them. For instance, Cheng and Pathak (42) reported that only 0 to 30% of first-stage GLH nymphs reached the adult stage on resistant varieties, whereas 76 to 90% of nymphs became adults on susceptible varieties. Increased ingestion by GLH from xylem vessels on resistant varieties may be due to the insect's effort to eliminate harmful chemicals from its body. In hemipterans, drinking helps maintain a sufficiently large water turnover for the removal of toxins, where possibly the bulk of water flows through the cuticle (43).

Conclusion

Practical implications of a full elucidation of biochemical bases of insect resistance in plants are tremendous. Identification of the chemicals that confer resistance or susceptibility, and study of their inheritance in crop plants, would greatly improve breeding for resistant varieties. If resistance involves more than one defense chemical, it may be possible to develop a relatively stable type of resistance since pests are not likely to overcome sensitivity to several substances simultaneously. It would also open new avenues for manipulation of insect behavior for use in pest management programs.

Proper understanding of the mechanisms of host plant resistance can lead to breeding varieties with long-term resistance. Major constraints to an interdisciplinary program for breeding pest resistant varieties include adequate analytical and bioassay facilities and the lack of formal working relationships among chemists, entomologists, and plant breeders. This is important since entomologists are not well versed in plant chemistry.

Literature Cited

1. Saxena, K. N. Ent. Exp. Appl. 1969, 12, 751-66.
2. Saxena, K. N.; Gandhi, J. R.; Saxena, R. C. Ent. Exp. Appl. 1974, 17, 303-18.
3. Saxena, R. C.; Pathak, M. D. In "Brown Planthopper: Threat to Rice Production in Asia"; International Rice Research Institute: Laguna, Philippines, 1979; pp. 304-17.

4. Whittaker, R. H. In "Chemical Ecology"; Sondheimer, E.;
    Simeone, J. B., Eds.; Academic: New York, 1970; pp. 43-70.
5. Reese, J. C.; Beck, S. D. Annals Entomol. Soc. Amer., 1976, 69,
    59-67.
6. Whittaker, R. H.; Feeny, P. P. Science 1971, 171, 757-70.
7. Reese, J. C. In "Host Plant Resistance to Insect Pests"; Hedin,
    P. A., Ed.; ACS SYMPOSIUM SERIES No. 62, American Chemical
    Society: Washington, D. C., 1977; pp. 309-30.
8. Waiss, A. C. Jr.; Chan, B. C.; Elliger, C. A. In "Host Plant
    Resistance to Insect Pests"; Hedin, P. A., Ed.; American
    Chemical Society: Washington, D. C., 1977; pp. 115-28.
9. Pathak, M. D.; Saxena, R. C. In "Breeding Plants Resistant to
    Insects"; Maxwell, F. G.; Jennings, P. R., Eds.; John Wiley &
    Sons: New York, 1980; Chap. 17.
10. Pathak, M. D. In "Concepts of Pest Management"; Rabb, R. L.;
    Guthrie, F. E., Eds.; North Carolina State University, 1970;
    pp. 138-57.
11. Nakano, K.; Abe, G.; Taketa, N.; Hirano, C. Jap. J. Appl.
    Entomol. Zool. (in Japanese with English summary), 1961, 5,
    17-27.
12. Sasamoto, K. Proc. Fac. Liberal Arts & Educ. Yamanasaki Univ.,
    Japan, No. 3, 1961, pp. 1-73.
13. Djamin, A.; Pathak, M. D. J. Econ. Entomol. 1967, 60, 347-51.
14. Panda, N.; Pradhan, B.; Samalo, A. P.; Prakasarao, P. S. Indian
    J. Agric. Sci., 1975, 45, 499-501.
15. Pathak, M. D.; Andres, F.; Calacgac, N.; Raros, R. "Resistance
    of Rice Varieties to Striped Rice Borers"; International Rice
    Research Institute, Tech. Bull. 11, 1971; p. 69.
16. Munakata, K.; Okamoto, D. In "Major Insect Pests of Rice
    Plant"; John Hopkins Press: London, New York, 1967; pp. 419-30.
17. Das, Y. T. Ent. Exp. Appl. 1976, 20, 131-34.
18. Saxena, R. C. "Biochemical Basis of Resistance in Crop Plants",
    Saturday Seminar, International Rice Research Institute: Laguna,
    Philippines, 11 March 1978, p. 25.
19. International Rice Research Institute (IRRI). In "Research
    Highlights for 1978"; IRRI: Laguna, Philippines, 1979; p. 8.
20. International Rice Research Institute (IRRI). In "Annual Report
    for 1981"; IRRI: Laguna, Philippines, 1983a; p. 74.
21. International Rice Research Institute (IRRI). In "Annual Report
    for 1982"; IRRI: Laguna, Philippines, 1983b; p. 65.
22. Pathak, M. D.; Cheng, C. H.; Fortuno, M. E. Nature (London),
    1969, 223, 502-4.
23. Sogawa, K.; Pathak, M. D. Appl. Ent. Zool. 1970, 5, 145-58.
24. Kim, M.; Koh, H.; Obata, T.; Fukami, H.; Ishii, S. Appl. Ent.
    Zool. 1976, 11, 53-7.
25. Yoshihara, T.; Sogawa, K.; Pathak, M. D.; Juliano, B. O.;
    Sakamura, S. Ent. Exp. Appl. 1979, 26, 314-22.
26. Yoshihara, T.; Sogawa, K.; Pathak, M. D.; Juliano, B. O.;
    Sakamura, S. Ent. Exp. Appl. 1980, 27, 149-55.
27. Buvat, R. "An Introduction to Plant Protoplasm"; McGraw-Hill
    Book Co.: New York, 1969; p. 256.
28. Shigematsu, V.; Murofushi, N.; Ito, K.; Kaneda, C.; Kawabe, S.;
    Takahashi, N. Agric. Biol. Chem. 1982, 46, 2877-96.
29. Kawabe, S.; Fukumorita, T.; Chino, M. Plant and Cell Physiol.
    1980, 21, 1319-27.

30. Saxena, R. C.; Puma, B. C. <u>Proc. 10th Annu. Res. Conf. of Pest Control Council of the Philippines</u>, Manila, 1979.
31. Saxena, R. C.; Okech, S. H. <u>J. Chem. Ecol.</u> 1985, 11 (in press).
32. Obata, T.; Koh, H.; Kim, M.; Fukami, H. <u>Appl. Ent. Zool.</u> 1983, 18, 161-9.
33. Saxena, R. C.; Barrion, A. A. <u>Korean J. Plant Prot.</u> 1983, 22, 52-66.
34. Gunther, E. "The Essential Oils", Van Nostrand: New York, 1952; Vol. V.
35. Robinson, T. "The Organic Constituents of Higher Plants: Their Chemistry and Interrelationships", Cordus Press: North Amherst, 1983; 5th ed.
36. Heinrichs, E. A.; Medrano, F. G.; Rapusas, H. R. "Genetic Evaluation for Insect Resistance in Rice"; International Rice Research Institute: Laguna, Philippines, 1985; p. 356.
37. Auclair, J. L.; Baldos, E.; Heinrichs, E. A. <u>Insect Sci. Appl.</u> 1982, 3, 29-34.
38. Khan, Z. R.; Saxena, R. C. <u>J. Econ. Entomol.</u> 1984, 77, 550-2.
39. Khan, Z. R.; Saxena, R. C. <u>J. Econ. Entomol.</u> 1985a, 78, 583-7.
40. Khan, Z. R.; Saxena, R. C. <u>J. Econ. Entomol.</u> 1985b, 78, 562-6.
41. McKey, D. In "Herbivores: Their Interaction with Secondary Plant Metabolites"; Rosenthal, G. A.; Janzen, D. H., Eds.; Academic Press: New York, 1979; pp. 55-133.
42. Cheng, C. H.; Pathak, M. D. <u>J. Econ. Entomol.</u> 1972, 65, 1148-53.
43. Stobbart, R. H.; Shaw, J. In "The Physiology of Insects"; Rockstein, M., Ed.; Academic: New York, 1974; Vol. V, pp. 361-446.

RECEIVED September 9, 1985

13

Potato Glandular Trichomes: A Physicochemical Defense Mechanism Against Insects

Peter Gregory[1,3], Ward M. Tingey[2], Dirk A. Ave[1], and Pierre Y. Bouthyette[1,4]

[1]Department of Plant Breeding and Biometry, Cornell University, Ithaca, NY 14853
[2]Department of Entomology, Cornell University, Ithaca, NY 14853

A wild potato species, Solanum berthaultii Hawkes, is resistant to insect pests by virtue of two types of glandular trichomes on its foliage. Type A is a short trichome with a four-lobed membrane-bound gland at its apex. Type B is a longer, simple trichome with an ovoid gland at its tip which continuously exudes a clear viscous exudate. Studies of this physicochemical resistance mechanism suggest the following series of events: (a) the insect lands upon the foliage and encounters Type B trichomes, (b) Type B trichomes coat the insect with a sticky exudate and, due to sesquiterpenoid action, agitate the insect, (c) the struggling insect disrupts the heads of Type A trichomes, and an oxidation reaction between a unique polyphenoloxidase enzyme (PPO) and phenolic substrate(s) results in quinone formation, (d) these events result in insect immobilisation, cessation of feeding and death. The relevance of this mechanism to breeding for improved potato resistance to insect pests is discussed.

Over the course of their association, plants have evolved many tactics to contend with herbivorous insects. Among the best known are chemical products of secondary metabolism which confer an adaptive advantage to the plant by acting as toxicants and repellents (1). No less important in protection from insect-induced stress is the plant's indumentum. Pubescence, once thought of only as a means for plants to minimize water loss, is now known to provide a defensive barrier against many species of herbivorous insects (2).

The types of pubescence active in defense are diverse, ranging from the simple, unbranched or stellate epidermal hairs

[3]Current address: International Potato Center (CIP), Apartado 5969, Lima, Peru.
[4]Current address: Allied Corporation, Solvay Research Laboratory, P.O. Box 6, Solvay, NY 13209.

0097-6156/86/0296-0160$06.00/0
© 1986 American Chemical Society

of maize, alfalfa, cotton, wheat, and soybean, to the striking secretory trichomes of solanaceous species (3). The glandular pubescence of potato defends against a broad complex of pest species, including aphids, leafhoppers, flea beetles, the Colorado potato beetle, the potato tubermoth complex, and spidermites (4,5,6,7). This wide spectrum of activity initiated considerable interest in the mechanistic basis for defense by glandular trichomes (8) and stimulated efforts to exploit these structures in potato breeding for pest resistance (9,10,11).

Hybrid potatoes bearing glandular trichomes from the wild Bolivian potato, Solanum berthaultii, are well defended against infestation and colonization by aphids, leafhoppers, and flea beetles (12). Populations of these damaging pests on elite hybrids are often reduced more than 85%, compared to those on susceptible commercial cultivars (13). Furthermore, many hybrids severely disrupt aphid feeding behavior and host acceptance in a manner comparable to that of their wild parent, an important attribute considering the worldwide importance of aphids as vectors of potato virus pathogens.

Potatoes bearing glandular trichomes also defend against the Colorado potato beetle, a devastating pest capable of total crop destruction (4). Although resistance will not eliminate the need for other control tactics in management of this pest, our field experience indicates that present levels of resistance in hybrids can substitute for as much as 40% of the current insecticide usage on susceptible cultivars (14).

Glandular Trichomes of S. berthaultii

Morphology. Resistance of S. berthaultii to insects is mainly due to the presence of two types of foliar glandular trichomes (Fig. 1). Type A is a short trichome (ca. 120 to 210 mµ in length) with a tetralobulate membrane-bound gland (50-70 mµ in diameter) at its apex. Type B is a longer (ca. 600 to 950 mµ in length), simple trichome with an ovoid gland at its tip which continuously discharges clear viscous exudate (the naked droplets at the tips are 20 to 60 mµ in diameter) (3). Type A trichomes are also present in S. tuberosum. In both species (and in other Solanum species bearing these trichomes), the morphology of Type A trichomes appears to be similar. Lyshede (15) found eight cells enclosing a secretion storage space in the trichome head. The secretion material consists of a lipophilic and polar phase. Release of secretory material occurs upon mechanical disruption of the head-stalk junction.

Biological Activity

Adhesive entrapment. Mortality and immobilization of aphids increases with a rise in density of Type A and Type B trichomes and with increased volume of Type A trichome glands. Insects landing on the foliage first encounter Type B trichome exudate which forms an adhesive coating on the tarsi. This accelerates the rupture of Type A trichome heads (6). Type A exudate is released onto the insect's body. The exudate rapidly becomes brown and hard. The insect becomes immobilized and feeding

Figure 1. A) Glandular trichomes on leaf surface of Solanum berthaultii (PI 310927). (X122). B) Foliar glandular and non-glandular trichomes of S. tuberosum cv. Katahdin. (X122). C) Viscous liquid exuding from tip of Type B trichome in S. berthaultii. (X250). Figures A and B reproduced from Tingey (12) with permission.

is disrupted (16,17).   If Type B trichomes are absent, as in
the case of S. tuberosum X S. berthaultii F$_1$ hybrids, fewer
Type A trichomes are ruptured and expression of resistance is
dramatically lowered.

Glandular trichomes of S. berthaultii provide the greatest
levels of resistance against immature life stages of pest species.
Encasement of feet and mouthparts of the green peach aphid by
trichome exudate is greatest on nymphs and least on adult aphids,
parallelling the trends in mortality (6).

Barrier effects and energetics of confrontation.   Glandular
trichomes may hinder the movement and normal activity of pest
species without trapping or disabling them.   This phenomenon
almost certainly plays a role in defense by S. berthaultii against
pests that seldom become trapped by trichome exudate (18), such
as the Colorado potato beetle.   Larvae of this species experience
reduced growth rates on S. berthaultii but rarely suffer
heightened mortality compared to those on the susceptible potato.
Larvae invest a significant portion of their activity in grooming
to remove trichome secretions from their tarsi and mouthparts.
This behavior detracts from time otherwise available for feeding.
In addition, the struggle to reach the plant surface through
a barrier of glandular trichomes probably entails energetic
costs that could lead to reduced growth and feeding rates.

Behavioral and sensory disturbance.   Potato glandular trichomes,
in addition to acting as a physical barrier to pests, manufacture
and/or store a profusion of plant metabolic products, some of
which profoundly influence insect behavior and metabolism (19,20).
The sesquiterpene components of potato trichome glands (21,22,23)
are potentially powerful semiochemicals and one of these,
E-β-farnesene, is well known for its ability to initiate evasive
behavior in aphids (24).   The dramatic alteration of aphid feeding
behavior on S. berthaultii reported by Lapointe and Tingey (25)
may be due to allomonal sesquiterpenes in trichome exudate and
will be discussed later.

Defensive Chemistry

Enzymatic browning.   Polyphenoloxidase (PPO) and peroxidase
(PO) activities have been found in S. berthaultii trichomes
(8).   The importance of browning in trichome-mediated defense
and the involvement of PPO and/or PO activity in browning are
shown by a combination of two lines of evidence.   First, treatment
of S. berthaultii foliage with cysteine, a phenolase inhibitor
(26), strongly inhibits the browning reaction and results in
greatly increased survival of aphids with corresponding lowered
levels of leg and mouthpart encasement by trichome exudate (Ryan,
unpublished data).   Cysteine has little or no direct effect on
the green peach aphid, as shown by controls involving aphid
survival measurements on cysteine-treated S. tuberosum foliage.
Second, within a population of S. tuberosum X S. berthaultii
F$_3$ hybrid clones, differing in resistance to aphids, infestation
levels varied by 8-fold; a 5-fold difference in glandular trichome
PPO activity was observed; and PPO activities were highly
correlated with aphid resistance (27).

The PPO and PO activities reported by Ryan et al (8) are due to action of two forms of PPO (α- and β-PPO) and a separate PO moiety (23,28). The α-PPO is an ortho-diphenol oxidase (O-DPO) containing 2 copper atoms with an estimated MW of 58,000. There are several isozymes of α-PPO each with a slightly different MW and PI. The β-PPO is also an O-DPO but with an estimated MW of 1.2×10^6. It contains 48 copper atoms and is bound to chlorogenic acid, the major endogenous phenolic compound in S. berthaultii trichomes. The β-PPO is probably a polymer consisting of approximately 24 α-PPO units each containing 2 copper atoms.

The presence of such a large PPO polymer appears to be unique to the potato trichomes. Even though polymers of PPO have been widely reported, the largest PPO observed in banana and apple leaves was estimated to be 150-180,000 (26). While aggregates of 2,4,8, and 16 subunits were observed in avocado and potato tubers (29,30), potato trichomes appeared to contain only two forms of PPO, a monomer of 58,000 and a polymer of 1,200,000, without intermediates. Interconversion between PPO aggregates, as found in apple fruit (31), was not achieved with potato trichome PPO. In several in vitro experiments in our laboratory, α-PPO did not polymerize even though a variety of substrate concentrations in the presence of α-PPO primer were used. Inhibition of endogenous phenol oxidation during trichome extraction with cysteine, dithiothreitol or ascorbic acid did not affect the levels of α- or β-PPO recovered.

In addition to being such a large polymer, β-PPO exhibited unusual kinetics. Data from Lineweaver-Burk plots showed that the Vmax of β-PPO (Vmax = 119 μmoles O_2 consumed/min/mg protein) was far greater than that of α-PPO (Vmax = 5.2 μmoles O_2 consumed/min/mg protein) while the Km values of α- and β-PPO for chlorogenic acid were similar (Km α = 0.84 mM; Km β = 1.37 mM) and were typical for plant catechol oxidases (Km ∿ 1mM (26)). The greater Vmax of β-PPO is possibly a polymerization-related synergistic effect. The presence of multiple active sites per se in a polymeric β-PPO would not explain the greater Vmax value because Vmax is calculated on an enzyme weight basis.

Isoelectric focusing of PPO's from chloroplasts, cytosol and trichomes of S. berthaultii leaves revealed another unusual enzymological feature; that is, the α-PPO is unique to the trichomes.

Sesquiterpenoid semiochemicals. The aphid-repellent effect of Type B trichomes of S. berthaultii (25) appears to be due to the presence of sesquiterpenes (22). Three major components, identified by GC-MS, were β-caryophyllene, β-cubebene and Δ-cadinene. E-β-farnesene was also identified, but was a minor component. GC-MS of the other major components in Type B trichome exudate indicated sesquiterpenoid structures, but these have not been identified.

The sesquiterpenoid fraction from S. berthaultii (PI 265858) foliage was qualitatively similar to that found in Type B exudate. Consequently, despite some quantitative differences in percentages of the individual sesquiterpenes, simple foliar extracts rather than Type B exudate provided the source of sesquiterpenes for bioassays.

As a control experiment surface washes from 200 S. tuberosum cv. Chippewa leaves showed the presence of β-caryophyllene and E-β-farnesene. Again, the sesquiterpene mixture for the bioassays was extracted from the foliage of cv. Chippewa and it contained β-caryophyllene, β-cubebene and E-β-farnesene (22). Other sesquiterpenes were present but have not yet been identified.

The effects of sesquiterpene mixtures from leaf extracts of wild and cultivated potatoes on the settling behavior of green peach aphid were studied in dual-choice assays (22). The sesquiterpene fraction of S. berthaultii had a strong repellent effect on aphid behavior. Due to use of the slow release medium Carboset, this effect was still apparent beyond six hours. The sesquiterpenoid fraction from cv. Chippewa leaves also strongly repelled aphids. The major foliar sesquiterpene in all extracts, β-caryophyllene, had no adverse effect on settling behavior.

E-β-farnesene, an aphid alarm pheromone, was present in cv. Chippewa extracts, and made up one fourth of the total sesquiterpene content. In comparison, the E-β-farnesene in the S. berthaultii extract made up only one hundredth of the total sesquiterpene content.

The sesquiterpenes of S. tuberosum foliage are not present in sufficient quantities at the leaf surface interface to discourage aphid settling. By contrast, the exudate of the Type B glandular trichomes on S. berthaultii foliage contains sesquiterpenes, thus enhancing the external level of these insect-repellent compounds.

Bromley and Anderson (32) showed with electrophysiological techniques that terpenes, including several sesquiterpenes, stimulate the primary rhinaria olfactory organs of the lettuce aphid, Nasonovia ribis-nigri. In addition, aphids generally perceive the alarm pheromone, E-β-farnesene, with the primary rhinaria (33,34). It should be noted, however, that stimulation of the same sensory organ by different compounds may lead to different behavioral responses. The typical alarm pheromone response is characterized by dislodgement or rapid movement away from the source. This alarm response is restricted to sesquiterpenes with structural features closely related to that of E-β-farnesene (35). In contrast, the sesquiterpenes of Type B glandular trichomes have elicited a generalized avoidance behavior. It may be possible to augment this repellent activity by selection of parental clones and hybrid potato clones bearing high levels of E-β-farnesene. Such clones, however, should also be selected for low β-caryophyllene levels because Dawson and coworkers (36) found that the alarm response of the green peach aphid elicited by E-β-farnesene was inhibited by β-caryophyllene, a major component of S. berthaultii trichomes and of S. tuberosum foliage. This effect of β-caryophyllene raises doubts about the claim of Gibson and Pickett (21) that E-β-farnesene of S. berthaultii trichomes is an important factor in resistance to aphids, because we found that trichomes of accessions studied by Gibson and Pickett (21) contain high concentrations of β-caryophyllene (22).

Conclusions and Outlook

Our present understanding of the glandular trichome-mediated
resistance mechanism in S. berthaultii can be summarized as
follows: (a) the insect lands upon the foliage and encounters
the tall Type B trichomes, (b) Type B exudate forms an adhesive
coating on the tarsi, and the sesquiterpenes in the exudate
elicit a disturbance behavior, (c) the insect struggles during
attempts to escape and breaks the Type A trichome heads, (d)
reaction of α- and/or β-PPO with phenolic substrate (chlorogenic
acid) occurs and oxidative processes are initiated, (e) quinones
are formed by enzymatic oxidation of phenols (the browning
reaction), (f) the insect becomes immobilized, ceases feeding,
and dies.
    Several questions remain unanswered. Is β-PPO with its
very high Vmax formed de novo from α-PPO upon rupture of Type
A trichome heads by the insect? Is the browning reaction
initiated by insect-induced damage, resulting in breakdown of
partitioning between PPO and substrate, and is action of a PPO
activator necessary? Does the insect contribute one or more
components to the browning reaction? Is the insect simply glued
in place by quinone-based brown polymer or do highly reactive
quinones chemically bind to the insect's surface? Have we only
just begun to unravel the intricacies of this elegant
physicochemical defense mechanism?

Acknowledgment

This work was supported, in part, by grant 7800454 and subsequent
grants from the Competitive Grants Office, SEA, USDA. A
publication of the Cornell University Agricultural Experiment
Station, NYS College of Agriculture and Life Sciences, a Statutory
College of S.U.N.Y.

Literature Cited

1. Rosenthal, G. A.; Janzen, D. H. Eds.; "Herbivores: Their Inter-
    action with Secondary Plant Metabolites"; Academic Press,
    New York, 1979; 718 pp.
2. Levin, D. A.; Q. Rev. Biol. 1973, 48, 3-15.
3. Tingey, W. M. In "The Leafhoppers and Planthoppers";
    Nault, L. R.; Rodriguez, J. G., Eds. Wiley and Sons: 1985.
    (In press).
4. Casagrande, R. J. Econ. Entomol. 1982, 75, 386-72.
5. Gibson, R. W. Ann. Appl. Biol. 1971, 68, 113-19.
6. Tingey, W. M.; Laubengayer, J. E. J. Econ. Entomol. 1981, 74,
    721-25.
7. Tingey, W. M.; Sinden, S. L. Amer. Potato J. 1982, 59, 95-106.
8. Ryan, J. D.; Gregory, P.; Tingey, W. M. Phytochem. 1982, 21,
    1885-1887.
9. Mehlenbacher, S. A.; Plaisted, R. L. In "Proceedings of the
    International Congress: Research for the Potato in the Year
    2000"; International Potato Center, Lima, Peru. 1983; p. 1228.
10. Mehlenbacher, S. A.; Plaisted, R. L.; Tingey, W. M. Amer.
    Potato J. 1983, 60, 699-708.

11. Mehlenbacher, S. A., Plaisted, R. L.; Tingey, W. M. Crop Sci. 1984, 224, 320.
12. Tingey, W. M. In "Advances in Potato Pest Management"; Lashomb, J. H.; Casagrande, R. Eds.; Hutchinson Ross Publ. Co., 1981; 288 pp.
13. Tingey, W. M.; Plaisted, R. L.; Laubengayer, J. E.; Mehlenbacher, S. A. Amer. Potato J. 1982, 59, 241-51.
14. Wright, R. J.; Dimock, M. E.; Tingey, W. M. ; Plaisted, R. L. J. Econ. Entomol. 1985, 77, (In press).
15. Lyshede, O. B. Ann. Bot. 1980, 46, 519-26.
16. Gibson, R. W.; Turner, R. H. PANS: Pest Articles and News Summaries 1977, 23, 272-7.
17. Tingey, W. M.; Gibson, R. W. J. Econ. Entomol. 1978, 71, 856-58.
18. Dimock, M. B.; Tingey, W. M. Mass. Agric. Exp. Sta. Bull. 1985. (In press).
19. Kelsey, R. G.; Reynolds, G. W.; Rodriguez, E. In "Biology and Chemistry of Plant Trichomes"; Rodriguez, E.; Healey, P. L.; Mehta, I., Eds. Plenum Press: New York. 1984, pp. 187-241.
20. Stipanovic, R. D.. In "Plant Resistance to Insect Pests"; Hedin, P. A. Ed.; ACS Symposium Series No. 208, American Chemical Society: Washington, D.C., 1983, pp. 69-100.
21. Gibson, R. W.; Pickett, J. A. Nature 1983, 302, 608-9.
22. Ave, D. A.; Gregory, P.; Tingey, W. M. Entomol. Exp. Appl. 1985 (Submitted).
23. Gregory, P.; Tingey, W. M.; Ave, D. A.; Bouthyette, P. Y. In "Insects and the Plant Surface"; Juniper, B. E.; Southwood, T. R. E. Eds. Edwin Arnold: London, 1985 (In press).
24. Nault, L. R.; Montgomery, M. E. In "Aphids as Virus Vectors"; Harris, K. F.; Maramorosch, K., Eds.; Academic Press, New York, 1977, pp. 527-545.
25. Lapointe, S. L.; Tingey, W. M. J. Econ. Entomol. 1984, 77, 386-89.
26. Mayer, A. M.; Harel, E. Phytochem., 1979, 18, 193-215.
27. Ryan, J. D.; Gregory, P.; Tingey, W. M. Amer. Potato J. 1983, 60, 861-68.
28. Bouthyette, P. Y.; Eannetta, N. T.; Hannigan, K. J.; Gregory, P. Phytochem. 1985 (Submitted).
29. Balasingam, K.; Ferdinand, W. Biochem. J. 1970, 118, 15.
30. Dizek, N. S.; Knapp, F. W. J. Food Sci. 1970, 35, 282.
31. Harel, E.; Mayer, A. M. Phytochem. 1968, 7, 199.
32. Bromley, A. K.; Anderson, M. Ent. Exp. Appl. 1982, 32, 101-10.
33. Nault, L. R.; Edwards, L. J.; Styer, W. E. Environ. Ent. 1973, 2, 101-5.
34. Wohlers, P.; Tjallingii, W. F. Ent. Exp. Appl. 1983, 33, 79-82.
35. Bowers, W. S.; Nishino, C.; Montgomery, M. E.; Nault, L. R. J. Insect Physiol. 1977, 23, 697-701.
36. Dawson, G. W.; Griffiths, D. C.; Pickett, J. A.; Smith, M. C.; Woodcock, C. M. Ent. Exp. Appl. 1984, 36, 197-99.

RECEIVED October 16, 1985

14

Genetic Control of a Biochemical Mechanism for Mite Resistance in Geraniums

R. Craig[1], R. O. Mumma[2], D. L. Gerhold[2,3], B. L. Winner[1,4], and R. Snetsinger[2]

[1]Department of Horticulture, Pennsylvania State University, University Park, PA 16802
[2]Department of Entomology, Pennsylvania State University, University Park, PA 16802

Host plant resistance to the two-spotted spider mite, Tetranychus urticae, was observed in the geranium, Pelargonium x hortorum. Initial studies indicated an association between leaf morphology and resistance; later studies concluded that resistance was due largely to the presence of glandular trichomes which produced a toxic exudate. The active components of this exudate have been identified as anacardic acids. Inheritance studies indicated that resistance was dominant to susceptibility and that two dominant complementary genes conditioned resistance.

Genetically conditioned pest resistance is important in efficient and economical production of floricultural plants. Resistance to insect and arthropod pests is usually genetically stable, and can be used to avoid plant problems associated with phytotoxic pesticides and residues. This allows for safe and economical production of plants without additional expense for chemicals and their application.

Prior to 1964, geraniums (Pelargonium x hortorum L. H. Bailey) were reported to have few two-spotted spider mite (Tetranychus urticae Koch.) injury problems. This paper describes a series of experiments conducted over a period of two decades which elucidate the genetic and causal mechanisms related to two-spotted mite resistance and susceptibility in geraniums. The paper is divided into six sections: 1) discovery of mite susceptibility; 2) assessment of genetic variation for resistance; 3) development of an efficient technique for evaluating resistance; 4) genetic studies; 5) studies on the mechanisms of resistance; and 6) identification of a specific biochemical metabolite which confers resistance. The divisions, for the most part, are chronological; however, for ease of presentation several experiments are grouped without respect to actual chronology.

[3]Current address: Department of Microbiology, University of Tennessee, Knoxville, TN 37996.
[4]Current address: Denholm Seeds, P.O. Box 1381, Lompoc, CA 93436

Discovery of Mite Susceptibility

In 1964, one of the authors (RC) observed a group of geraniums in
the horticultural greenhouse at The Pennsylvania State Univesity
which were chlorotic and less healthy than surrounding plants. Upon
inspection by R. Snetsinger the plants were observed to be infested
by two-spotted spider mites. Plants which were immediately adjacent
were not infested, and further inspection indicated that infested
plants were randomly located throughout the greenhouse. Since the
pedigrees of the plants were known, it was possible to verify that
all susceptible plants had resulted from a single cross - 'King
Midas' and a dark orange-red flowered selection from 'Florists'
Mix', designated G71. 'King Midas' had resulted from x-ray treatment
and had been selected because of its clear, orange flower color (1).
It was assumed, and later proven, that 'King Midas' contributed
genetic factors for mite susceptibility.

Assessment of Genetic Variability for Resistance

The discovery of susceptibility where it previously had not been
reported stimulated us to conduct a study of variation within
Pelargonium. Snetsinger et al. (2) studied the relative resistance
of seven species of Pelargonium and six accessions of P. x hortorum.
They noted that four of seven species (P. capitatum, P. australe, P.
tomentosum and P. rapaceum) were more resistant than the other three
(P. fulgidum, P. odoratissimum and P. graveolens). Of the
accessions of P. x hortorum which were evaluated, three were commercial cultivars (two diploids and one tetraploid) and the other three
were 'King Midas', G71 and their hybrid. The commercial cultivars
were resistant or moderately susceptible when evaluated on the basis
of the number of days that mites survived and number of eggs
oviposited on detached whole leaves. Of the genetic accessions,
'King Midas' was rated as susceptible and G71 and the hybrids were
rated as resistant. It is important to note that none of the
species are related to P. x hortorum.
     Recent work by Potter and Anderson (3) on P. peltatum, the
ivy-leaved geranium, indicated that genetic variability existed
among cultivars for two-spotted spider mite resistance.
Observations by one of the authors (RC) further indicated that P. x
hortorum tetraploid cultivars of European origin exhibit
differential degrees of mite susceptibility under greenhouse
conditions.

Development of An Efficient Evaluation Technique

In order to effectively evaluate resistance and susceptibility, an
efficient evaluation technique was required. MacDonald et al. (4)
tested four evaluation techniques: visual rating of whole plants, an
attached whole leaf method, a detached whole leaf method and a leaf
disc method. The leaf disc method was considered the most
satisfactory since it was space efficient, it allowed for
replication both within and between leaves of an individual plant,
it caused no harm to the plants and it yielded highly reproducible
results.

The method consisted of cutting a 12 mm disc from the leaf with a cork borer placed against a clean block of wood. The surface of the leaf to be evaluated was placed away from the wooden block. Individual discs were placed on filter paper in 35/60 mm microdiffusion dishes. Water was added to moisten the filter paper to prevent leaf disc dessication and to prevent migration of the mites. One adult female mite from an isogenic culture maintained on bean leaves was placed on each leaf disc. The dishes were put on trays and placed in a growth chamber maintained at 23.3° C during the 12-hour photoperiod and at 17.6° C during the 12-hour nyctoperiod. Resistance was evaluated by measuring fecundity and survival during a 72-hour period. This method requires sufficient replication to establish valid resistance values.

Genetic Studies of Resistance

Genetic studies were divided into two parts, some early work by Snetsinger et al. (2) and a more extensive study by Winner (5). The first assessment of genetic variation for mite resistance in Pelargonium, demonstrated the susceptibility of 'King Midas', G71, and their hybrid. Snetsinger et al. (2) concluded that the observed resistance was a dominant trait. They also studied nine F_2 progeny (self-pollinated progeny of the hybrid) and observed differential levels of susceptibility. Although no conclusions could be drawn from such a small sample, it appeared that resistance might be simply inherited (2 plants were highly susceptible, 2 plants were moderately susceptible and 4 plants were resistant).

The small population resulted from several difficulties which were inherent in 'King Midas'; it was pollen sterile and the plant had low vigor. It was necessary to identify a susceptible inbred line which could be used in genetic studies. A survey of available genetic material resulted in the acquisition of G60, a pink flowered mite-susceptible line. In addition, the survey indicated that a highly resistant line, G54, satisfied the criteria for genetic studies; it was used to replace G71, the original resistant parent.

Winner (5) conducted a genetic study of mite resistance using parental (G54 and G60), F_1, F_2 and backcross generations. He evaluated 2115 plants by the leaf disc method (9000 leaf discs) as described earlier. Each plant was evaluated for mite fecundity, which was measured by the number of eggs laid in a 72-hour period, and mite survival at 72 hours. A resistance index, which was the product of fecundity times survival, was calculated for each plant. Lower indices indicated resistance (few eggs and low survival) and high indices indicated susceptibility. He used between four and eight replications of each plant to establish resistance; resistance indices ranged from 0-197 for the resistant parent and from 142-475 for the susceptible parent. The resistance index was used to classify each plant as resistant or susceptible. Bimodal distributions of resistant and susceptible phenotypes were observed in the segregating generations; a resistance index of 180 was used as the discrimination point for resistance versus susceptible progeny.

Winner (5) concluded that the resistant and susceptible parents were true breeding for their respective phenotypes and that inheritance was conditioned by nuclear genes. Resistance was completely dominant to susceptibility and was conditioned by two dominant complementary loci which he designated as $\underline{ts}_1$ and $\underline{ts}_2$. This model results in an expected ratio of 9 resistant: 7 susceptible progeny in the F_2 generation. One additional observation was that it is possible to have true-breeding heterozygotes. Since both loci must be dominant for the expression of resistance, susceptible pheonotypes could be heterozygous at either, but not both, loci.

Mechanisms of Resistance

Two experimental areas with respect to mechanisms of resistance to the two-spotted spider mite were studied. Chang et al. (6) investigated anatomical and morphological characteristics of resistant and susceptible phenotypes and Stark (7) investigated biochemical mechanisms. Chang's studies were conducted prior to the genetic research and Stark's experiments were conducted concurrently and after the genetic research.

Morphological-Anatomical Studies. Chang studied the leaf surface characteristics of 'King Midas' (G10), the original susceptible parent, and G54 the new resistant parent. She also sectioned leaves to determine the thickness of the cuticle and epidermal cells of each phenotype. Prior to conducting her studies she established the resistance of each phenotype by the leaf disc technique. The results are best illustrated by the number of eggs oviposited on the adaxial and abaxial surfaces:

	Adaxial	Abaxial
G10 (S)	2.67	3.04
G54 (R)	1.62	1.33

Fewer eggs were observed on the resistant parent than on the susceptible parent and fewer eggs were observed on the adaxial versus the abaxial leaf surfaces. Most species of two-spotted mites generally feed and oviposit on the abaxial (lower) leaf surfaces (8).

Based on her mophological studies, Chang concluded that there was no difference in the total number of trichomes present on resistant and susceptible plants; however, the resistant phenotype had more glandular trichomes on both leaf surfaces than the susceptible phenotype. Anatomically, the resistant parent had a thicker leaf cuticle and thicker epidermal cells than the susceptible parent. She related resistance to the differences in thickness of the cuticle and epidermal cells and suggested that they influenced feeding behavior of the mite, vis-a-vis stylet length.

Biochemical Studies. Stark (7) confirmed the findings of Chang with respect to trichome number. In addition, with the use of the scanning electron microscope, he was able to determine that resistant plants had four types of trichomes, two were glandular (50μ and 100μ in length) and two were spine-shaped (100-300μ and 700μ). Susceptible plants had both types of glandular trichomes but only the shorter spine-shaped trichomes. The most noticeable difference between resistant and susceptible plants was the presence of an exudate on the glandular trichomes of resistant plants. Susceptible plants did not produce an exudate. The exudate had both toxic and adhesive effects on adult mites and adversely affected oviposition:

| | Percent Mites | | Number |
	Adhered	Mortality	Eggs
Resistant	39	94	.23
Susceptible	0	24	5.00

Stark collected the pure exudate with small capillary tubes and topically applied it to mites, mite eggs, quiescent pre-adult whitefly, and potato aphid nymphs. In all cases a significant degree of mortality was observed.

Identification of A Specific Biochemical Which Confers Resistance

Gerhold et al. (9) developed research techniques to identify the biochemical metabolite which conditioned resistance. They used the same plant materials that had been included in Winner's genetic studies and Stark's research. All plants were grown from seed and were maintained without the use of pesticides. Included in their study was a resistant parent, a homozygous recessive susceptible parent, and their F_1 hybrid. In addition, they tested a susceptible parent which had one dominant locus for resistance. Basically their technique involved the collection of exudate and subsequent analysis by gas- and high-performance-liquid chromatography, mass spectrometry, and nuclear magnetic resonance (NMR) spectroscopy.

They developed an efficient technique for the collection of relatively pure exudate. By just touching the leaf with a clean microscope slide, drops of exudate were deposited on the slide. The slide was rinsed with ethyl acetate to produce an extract of the exudate.

A bioassay was developed to demonstrate that the exudate extract did exhibit acaricidal properties. A sandwich of two microscope slides and a teflon spacer with a hole in the center was used as the arena. Trichome extract (226μ g) and 15μ l of 5% Tween 20 were placed on a filter paper disc (12 mm) and six adult mites were bioassayed for mortality at 72 hours with the following result:

	% Mortality
Exudate	75
Control	14

The actual difference of 61% compares favorably to the results of Stark's bioassay (70%) with the leaf disc technique.

Gas-liquid chromatograpy (GLC) of the exudate produced two peaks designated A and B in a 35:65 ratio. Mass spectra of the two peaks gave molecular ions of 302 and 330 m/z. The fragmentation pattern of the two peaks was essentially identical except for the differences in the molecular ions. High resolution mass spectra of the GLC peaks yielded empirical formulas of $C_{21}H_{34}O$ and $C_{23}H_{38}O$ demonstrating that peak A and B differed by a C_2H_4 unit.

The exudate was shown to absorb strongly in the ultraviolet region; the UV absorption was affected by base which implied the presence of an ionizable proton. These data suggested that the molecules were aromatic.

A high-performance liquid chromatography (HPLC) procedure was developed to analyze the exudate since the constituents absorbed strongly in the UV. Reverse phase chromatography (μ Bondapak C_{18} column with a 90% acetonitrile-10% 0.1 N acetic acid-water mobile phase) resulted in the separation of two peaks designated A' and B'.

Mass spectra were determined for the crude extract and for compounds A' and B' from the HPLC analysis. The solid-inlet probe mass spectra did not exactly correspond to the mass spectra of peaks A and B from the GLC analysis. Peaks A' and B' gave molecular ions of 346 and 374 m/z, respectively, exactly 44 units higher than peaks A and B. High resolution mass spectrometry showed the difference to be CO_2. Thus, these data implied that compounds A' and B' were undergoing decarboxylation upon gas chromatography and that the resulting peaks A and B were decomposition products of A' and B'.

Interpretation of the mass spectra confirmed the presence of unsaturated hydrocarbon chains of $C_{15}H_{29}$ (A') and $C_{17}H_{33}$ (B') presumably connected to an aromatic nucleus possessing a carboxyl group and phenolic group. A search of the literature indicated that anacardic acids fit the description of compounds A' and B' and they were known to be heat labile resulting in the loss of a carboxyl group (10). Anacardic acids have been previously reported as occuring in the family <u>Anacardiaceae</u> (11). The proposed structures of compounds A' and B' are illustrated below and the mass spectral fragmentation pattern for B' is shown in Figure 1.

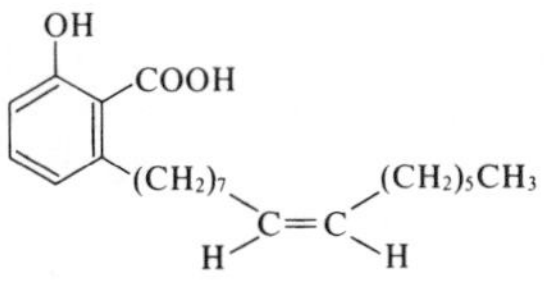

Compound A'
(Romanicardic Acid)

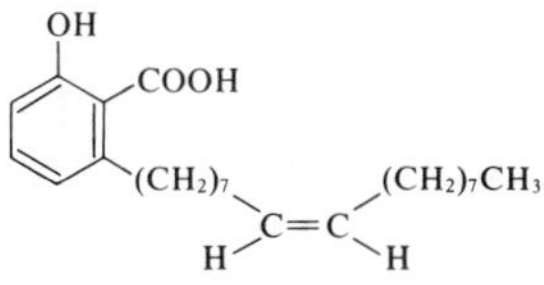

Compound B'
(Geranicardic Acid)

NMR spectroscopy was utilized to confirm the structure of compounds A' and B'. Each type of proton was clearly detectable and the analysis demonstrated the presence of three aromatic protons, two vinylic protons, two benzylic protons, three methyl protons, and numerous methylene protons (Table I).

Table I. NMR spectrum of geranium trichome exudate.

δ(in $CDCL_3$)	# of Protons	Splitting Pattern	Probable Group
.90	3	triplet (complex)	CH_3-C-
1.26	~25	multiplet	$C-CH_2-C$
1.60	~3	multiplet	$CH_2-C-\phi$
2.01	4	multiplet	$-CH_2-C=C$
2.97	2	triplet J=7.31, 7.31	$-CH_2-\phi$
5.35	~2	triplet J=4.45, 4.73	$-CH{=}CH-(\underline{cis})$
6.77	1	doublet J=7.74	ϕ(#5 position)
6.87	1	doublet J=8.17	ϕ(#3 position)
7.36	1	triplet J=8.17, 7.74	ϕ(#4 position)
11.00	2	singlet (broad peak)	$HO-\phi$ and $HOOC-\phi$

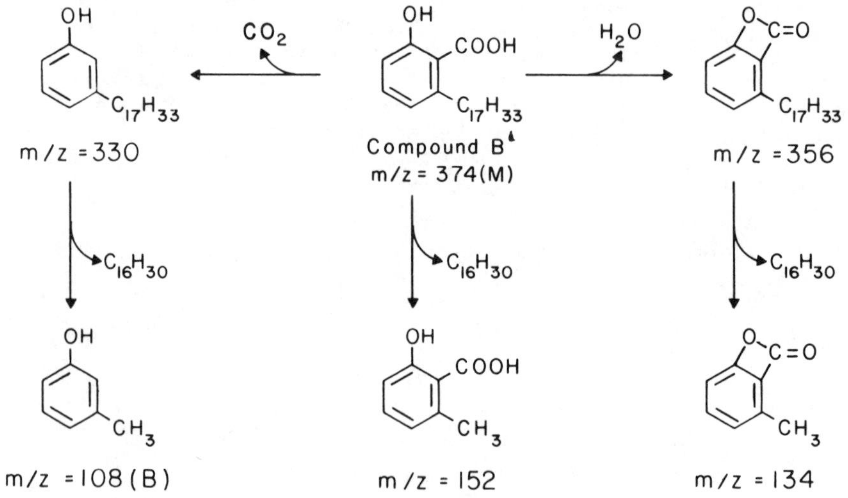

Figure 1. Mass spectral fragmentation scheme of compound B'.
Reproduced with permission from Ref. 9. Copyright 1984 Plenum
Publishing Corp.

In addition, NMR confirmed the presence of a cis double bond. Collectively these data correspond to romanicardic and geranicardic acids for A' and B', respectively. It should be noted that the double bond placement is based on the biogenic relationship to fatty acid biosynthesis and may actually be located at a different position (12). Future chemical studies will be required to establish the position of the double bonds.

To verify the association between genetic and biochemical parameters, Gerhold et al. analyzed the exudates from four genotypes which had been designated by Winner (5).

Plant	Compounds	Genotype
Resistant Parent	A' and B'	$Ts_1Ts_1Ts_2Ts_2$
F_1 Hybrid	A' and B'	$Ts_1ts_1Ts_2ts_2$
Susceptible Parent	None	$ts_1ts_1ts_2ts_2$
Susceptible Parent	None	$Ts_1ts_1ts_2ts_2$

They concluded that both loci must exist in the dominant condition for the production of exudate and the anacardic acids. No exclusive acaricidal action of A' and B' could be assigned.

The Future

Five areas are suggested for future research. It is important to develop a better understanding of the genetic mechanism at the biochemical level. What is the action of the two loci which are involved in mite resistance? This will require detailed knowledge of the biosynthesis of romanicardic and geranicardic acids, particularly enzymological considerations. Secondly the site of biosynthesis must be determined; this can be determined from an investigation of the morphological and anatomical structure of leaves from resistant plants. Of particular importance will be glandular trichomes and associated epidermal cells.

The third area involves the production of the compounds in tissue culture. This research would be important both for basic studies of the exudate and for practical considerations involved in large scale production of the compounds. The fourth research focus is more long range. It might be possible to isolate or synthesize the genes which condition mite resistance. These genes could be cloned and inserted into other organisms - both higher plants and microorganisms. Two outcomes are possible - first the transfer of mite resistance to plants which are now susceptible; and second, the production of the biochemicals for pesticide use.

The final area of study is related to plant breeding. Many improved cultivars of geraniums are being developed by university and commercial plant breeders. It is critical that new types be resistant to the two-spotted spider mite; thus the techniques and information which have been elucidated in our research can assist breeders in their cultivar development programs.

Summary

We have shown that genetically controlled biochemical compounds in geraniums confer resistance to the two-spotted spider mite. Initial discovery of mite susceptibility among breeding materials led to an assessment of variation both at the genus and species levels. This

early study indicated that a standardized technique was necessary
for research on the inheritance of mite resistance. A leaf disc
technique was developed and used to study the genetic basis of
resistance. Two dominant complementary genes condition resistance.
Both morphological and biochemical mechanisms were associated with
resistance. Resistant plants were shown to have thicker cuticles
and epidermal cells, and to have long and short spine-shaped
trichomes along with long and short glandular trichomes. The
glandular trichomes of resistant plants produce an exudate which has
both adhesive and toxic effects on mites. The exudate contains two
compounds - romanicardic and geranicardic acids. The structure of
these compounds has been elucidated by HPLC, GLC, NMR and mass
spectrometry analysis. Future work is planned on the enzymology and
biosynthesis of these compounds.

Acknowledgment
Paper No. 7234 in the Journal Series of the Pennsylvania
Agricultural Experiment Station. The authors acknowledge the
contributions of the numerous technicians and graduate students who
were involved in this research.

Literature Cited

1. Richards, E. Geraniums Around the World 1958. 6(1), 11.
2. Snetsinger, R.; Balderston, C. P.; Craig, R. J. Econ. Ent.
    1966, 59, 76-8.
3. Potter, D. A.; Anderson, R. G. J. Am. Soc. Hort. Sci. 1982,
    107, 1089-92.
4. MacDonald, A. J.; Root, J.; Snetsinger, R.; Craig, R.
    Melsheimer Ent. Series 1971, 8, 1-4.
5. Winner, B. L. Inheritance of Resistance of the Two-spotted
    Spider Mite, Tetranychus urticae Koch in the Geranium,
    Pelargonium x hortorum Bailey. M.S. Thesis, The Pennsylvania
    State University, University Park, Pennsylvania. 1975.
6. Chang, K. P.; Snetsinger, R.; Craig, R. Entomol. News 1972,
    83, 191-7.
7. Stark, R. S. Morphological and Biochemical Factors Relating to
    Spider Mite Resistance in the Geranium. Ph.D. Thesis, The
    Pennsylvania State University, University Park, Pennsylvania.
    1975.
8. Eyo, B. A.; Snetsinger, R.; Craig, R. Melsheimer Ent. Series
    1974, 14, 1-5.
9. Gerhold, D. L.; Craig, R.; Mumma, R. O. J. Chem. Ecol. 1984,
    10, 713-21.
10. Royals, E. E. In "Advanced Organic Chemistry"; Prentice-Hall,
    New York. 1954; p. 108.
11. Wasserman, D.; Dawson, C. R. Ind. Eng. Chem. 1945, 37, 396-9.
12. Geissman, T. A. In "Biogenesis of Natural Compounds";
    Bernfeld, P., Ed.; Pergamon Press, Oxford, England. 1963; p.
    574.

RECEIVED October 10, 1985

ALLELOCHEMICALS AS PEST CONTROL AGENTS

15

Secondary Metabolites from Higher Plants
Their Possible Role as Biological Control Agents

Horace G. Cutler[1], Ray F. Severson[1], Patsy D. Cole[1], D. Michael Jackson[2], and Albert W. Johnson[3]

[1] Richard B. Russell Center, USDA, Agricultural Research Service, Athens, GA 30613
[2] Tobacco Research Laboratory, USDA, Agricultural Research Service, Oxford, NC 27565
[3] Pee Dee Research and Education Center, Clemson University, Florence, SC 29503

The leaf surface chemicals of green tobacco (Nicotiana tabacum L.), derived mostly from the trichomes, consist primarily of α and β -4, 8, 13-duvatriene-1, 3-diols, α and β -4, 8, 13-duvatrienols, (DVT diols, ols), docosanol, hydrocarbons and sucrose esters. Of these, the DVT diols, DVT ols, and sucrose esters, significantly inhibit the growth of wheat coleoptiles. In addition, DVT diols and ols are ovipositional stimulators for the tobacco budworm Heliothis virescens and experiments show that egg numbers and DVT concentration are logarithmically proportional. Gram positive bacteria, Bacillus subtilis and B. cereus, are inhibited by α-DVT diol:gram negative bacteria and the fungi Curvularia lunata and Aspergillus flavus are not inhibited. Sucrose esters, wherein the acyl groups are mixtures of C_{2-8} fatty acids, also inhibit growth of wheat coleoptiles, but the 3-methylvalerate sucrose ester is the most inhibitory. The 3-methylvalerate sucrose ester also inhibits gram positive B. subtilis, B. cereus, and Mycobacterium theromosphactum, but does not inhibit gram negative bacteria, or the fungi C. lunata and A. flavus. Neither 3-methylvalerate glucose, nor methylvaleric acid, nor the K^+ salt have biological activity. Combinations of high levels of DVT diols, and sucrose esters, inhibit budworm larval development.

Earlier, we described the isolation and identification of two isomeric diterpenes from the leaves of young tobacco plants (Nicotiana tabacum L., cv. Hicks) that significantly (P<0.01) inhibited the growth of etiolated wheat coleoptiles (Triticum aestivum L., cv. Wakeland) in concentrations that ranged from 10^{-3} to 10^{-5} M (1,2). X-ray crystallography confirmed the structures

0097–6156/86/0296–0178$06.00/0
© 1986 American Chemical Society

as α and β - 4, 8, 13- duvatriene -1,3-diol (DVT diols) (Figure 1), $C_{20}H_{34}O_2$ (3), members of the cembrene family originally isolated as potential flavor components by Roberts and Rowland (4). Our experimental data had suggested that young upper-most leaves control the growth of lateral buds in tobacco by producing natural growth inhibitors and, furthermore, because β-DVT diol inhibited wheat coleoptiles 15% at 10^{-5} M we suspected that it was the bud dormancy controlling substance. At the time that these experiments were carried out, the etiolated wheat coleoptile bioassay was used almost exclusively for detecting either plant growth promoting or inhibiting substances. The α-DVT diol did not inhibit wheat coleoptiles as much as did abscisic acid (ABA), the compound responsible for axillary bud dormancy in many species (5), which inhibited coleoptiles 73% at 10^{-5} M. However, we thought that α and β -DVT diols may be tissue specific for tobacco; that is, they exerted a greater influence on the axillary buds of tobacco. This line of reasoning was reinforced because neither ABA was found in Hicks tissue, nor did exogenously applied (±) ABA control the growth of axillary buds. We further suggested that α and β -DVT diols may play an interesting biological role (6) and other studies (7) supported that statement. The ciliated organism, Tetrahymena pyriformis, when treated with aqueous solutions of β -DVT became immobile. As the concentration of DVT increased, cilia movement became proportionally slower until there was complete cessation and the organism died.

Other diterpenes inhibit the growth of wheat coleoptiles, specifically, cis-abienol from N. tabacum cv. PB (Bergerac); 2-hydroxymanool, the eutectic mixture of sclareol and epi-sclareol, pure sclareol, manool, manoyl oxide (active at 10^{-3} M) and α-levantenolide (the β -isomer was inactive) from N. glutinosa. Labdanediol, isolated from Turkish tobacco, is also quite active (6). Following these discoveries, further work on the plant growth regulatory properties of the diterpenes ceased until recently when the confluence of experimental results obtained from studies with the tobacco budworm, Heliothis virescens (F.), and our interest in other potential biological properties of diterpenes, and other natural products from the tobacco plant, occurred. It was noted, in certain field trials conducted at Tifton, Georgia, that different tobacco cultivars planted in adjacent rows were only slightly damaged by H. virescens and tobacco hornworm, Manduca sexta (L.), larvae while neighboring rows were decimated by larvae. In those rows, the parenchyma and spongy mesophyll had been completely eaten leaving only veins, mid-ribs, and stalks standing. Obviously, there were either feeding repellants or insect inhibitors in the resistant variety and, possibly, attractants in the susceptible cultivars.

We now report the diverse biological activity of the tobacco cuticular DVTs, the sucrose esters and the implications of the presence of these compounds in the plant-insect-microorganism ecosystem.

The source of these materials in tobacco is the leaf and stem trichomes. Compounds exude from the glandular trichomes, the morphology of which differ from cultivar to cultivar. For example, tobaccos with simple, or nonglanded trichomes produce only cuticular hydrocarbons. The profusion of the trichome

exudates makes the leaf and stem sticky to the touch so that
handling plants, or merely brushing against them in the field,
quickly leads to the deposition of a sticky clear 'gum' on the
hands. The gum is composed of diterpenoids, long chain alcohols,
wax esters and sucrose esters (8). It is important to note that
the glandular exudates are not secreted as a bolus, or one time
event, but are produced continuously throughout the growing cycle
of the plant. As we shall see later, this may be important to
understanding certain biological events. In addition, both the
quantities and types of cuticular products may vary from one
tobacco species and cultivar to another (8).

Cuticular leaf extracts used in the experiments were made by
cutting the terminal 45 cm from 6-week-old field-grown tobacco
plants when they were approximately 110 cm tall. Two 4 L beakers,
each containing 2 L of CH_2Cl_2, were placed side by side and plant
tops were dipped four times, for approximately 2 seconds per dip,
in each beaker. Generally, 50 plants of each tobacco cultivar
were extracted. The extract was filtered through anhydrous sodium
sulfate on filter paper and the solvent was evaporated at 40°C,
under vacuum. Approximately 3 grams of leaf surface compounds
were realized by this method which extracts at least 96%+ of the
cuticular exudates. Next, the material was solvent partitioned
between 150 mL each of hexane: 80% methanol in water. It was
determined, by GC analyses, that in the general schema the hexane
fraction contained hydrocarbons, DVT ols, cis-abienol, fatty
alcohols, wax esters and other non-polar terpenes. The polar
phase contained DVT diols, labdenediol, sucrose esters and other
polar terpenes. However, not all cultivars contain these
substances and five different tobaccos were extracted for
component isolation. Further fractionation of the α and β -DVT
diols, labdenediol and sucrose esters was effected on Sephadex LH
20 columns (2.5 x 44 cm) using a mobile phase of $CHCl_3$ containing
exactly 0.75% ethanol at a 2 mL/minute flow rate to yield isolates
of these compounds. Additional clean up was either by
recrystallization, or column chromatography through basic alumina,
to give pure compounds. Detailed information regarding extraction
methods and purification has been reported (9, 10, 11). The
mixture of sucrose esters was resolved by reverse-phase
chromatography using C_{18} and acetonitrile: water (50:50).

Analyses of the cuticular components from leaves were
accomplished by glass capillary gas chromatography (8). Cuticular
compounds from each tobacco cultivar, representing 0.3-0.6 grams
green leaf tissue weight, or 10-20 cm² leaf area, were placed in
test tubes with an internal standard of heptadecanol (50 ug) and
the solvent was removed under a stream of nitogen at 40°C. Next,
50 uL of BSTFA:DMF (1:1) was added, the tube was sealed and heated
for 30 minutes at 76°C. Upon cooling, 50 uL of BSA-pyridine (1:1)
was added to prevent hydrocarbon precipitation. Samples were
analysed on a Hewlett-Packard 5840 GC using a 0.3 mm i.d. x 25 m
thin film SE-50 fused silica WCOT column with H_2 carrier gas and a
flame ionization detector (8).

Bioassays were divided into three parts: plant, microbial
and insect. The insect bioassays included ovipositional and larval
development studies. The plant assays consisted of 4 mm
coleoptile segments, cut from the apices of etiolated 4-day-old

wheat seedlings (<u>Triticum aestivum</u> L., cv. Wakeland), which were
floated in phosphate-citrate buffer at pH 5.6 supplemented with 2%
sucrose (<u>12</u>) and included the natural products to be tested.
Control tubes contained buffer-sucrose solutions only. Sections
were incubated for 18 hours at 21.7°C, measured (<u>13</u>) and data were
analyzed by a multiple comparison procedure (<u>14</u>). Details
concerning the bioassay and the limits of detection with respect
to types of compounds have been published (<u>13</u>, <u>15</u>).

Microbial inhibition bioassays consisted of solid media in
petri dishes densly streaked with the organism to be challenged.
The bacteria, <u>Bacillus subtilis</u> (gram positive) and <u>Escherichia</u>
<u>coli</u> (gram negative) were cultured on diognostic sensitivity test
agar (DST-Oxoid) at 37°C overnight with 4 mm disks impregnated
with the compounds to be assayed. The compounds were dissolved in
either acetone, or dimethylsulfoxide, then added to the disks at
selected concentrations. Bacterial assays were also run with
selected compounds against <u>B. cereus</u> (gram positive),
<u>Mycobacterium thermosphactum</u> (gram positive), <u>E. cloacae</u> (gram
negative) and <u>Citrobacter freundii</u> (gram negative) on DST. Fungi
were grown on potato-dextrose agar (Difco) and treated 4 mm disks
were added to the heavily seeded medium, then incubated for 3-5
days, depending upon the genus and species of the fungi used. The
two fungi treated were <u>Curvularia lunata</u> and <u>Aspergillus flavus</u>,
the latter a producer of the aflatoxins.

Ovipositional bioassays were conducted in rectagonal screened
cages 3.7 x 3.5 x 2.1 m (high), though these were later modified
to 2.4 x 2.4 x 2.0 m (high) because of costs. Treatments
consisted of placing 5- to 7-week-old pot grown tobacco plants in
groups of 4 at opposite corners within the screened cages. Ten
female <u>H. virescens</u> moths, which had been kept with 10 males in
mating cages following hatch, were released in the late afternoon
and the following morning all budworm eggs were counted. When at
least 50 eggs were discovered, per cage, the data were used and,
additionally, if at least 5 moths in each cage were determined to
have mated as evidenced by spermatophore examination.

Applications of isolated compounds for ovipositional
preference studies were made to upper and lower leaf surfaces in
aerosol using a small air-brush. Compounds were dissolved in the
carrier solution formulated with distilled water and acetone
(1:3). Control plants were treated with carrier solution only
(<u>17</u>).

<u>Ovipositional properties of DVT diols and DVT ols</u>. In 1978, it
was proposed that a tobacco cultivar, Tobacco Introduction (TI)
1112, was resistant to tobacco budworm because it lacked certain
trichome exudates (<u>16</u>). Also, there were fewer budworm eggs on TI
1112 leaves compared to egg masses found on two other flue-cured
cultivars. Four years later, Johnson and Chaplin (<u>17</u>) reported
that trichome secretions influenced budworm moth oviposition in
certain cultivars and, furthermore, resistant cultivars were
capable of producing antibiotic substances that affected first
instar larvae. Later, the leaf surface chemistry of TI 1112 was
studied in detail and compared with that of budworm susceptible NC
2326 (<u>18</u>). The cuticular substances of NC 2326 were composed of
high levels of α and β-DVT diols, a series of C_{25} to C_{30} saturated

hydrocarbons, lower amounts of C_{18} to C_{30} fatty alcohols and C_{30} to C_{52} wax esters. The predominant long chain alcohols were docosanol, a compound thought to have plant growth regulating properties (19) and eicosanol (Figure 2a). In comparison, the cuticular compounds of TI 1112 contained only traces of α and β-DVT diols, fatty alcohols, or wax esters. But there were similar concentrations of hydrocarbons (Figure 2b). A series of experiments ensued over a three year period in which the whole leaf wash of NC 2326 cuticular extracts were partitioned against hexane and methanol:water (vide supra) and each was applied to the resistant TI 1112 (18). The treated plants were then challenged with moths in caged experiments. Capillary GC showed that the methanol:water solubles consisted entirely of α and β-DVT diols, and other polar diol degradation products, while the hexane soluble portion contained traces of α and β-DVT diols and ols, alcohols, and hydrocarbons.

A synopsis of the experiments (Table I) shows that on untreated TI 1112 plants 6,610 eggs were laid overnight by tobacco budworm moths while 21,424 were deposited on NC 2326. In the wild, it is more difficult to find budworm eggs on TI 1112 plants and two reasons are attributed to the apparent anomolies in these experiments. First, the flying pattern for the moths is influenced by the size of the cages, and in earlier experiments the figures for eggs on TI 1112 were lower in bigger cages. Second, moths reared in captivity appear to undergo some unexplained alterations, perhaps due to different selection pressures in laboratory colonies. Nevertheless, there is significant (P<0.01) difference in the number of eggs laid on the leaves of each cultivar.

When α and β-DVT diols were applied to leaves of TI 1112 and the plants challenged with female moths, only 118 eggs were deposited on control plants while a significant (P<0.05) number, 2,104 were laid on treated plants (Table II).

Perhaps the most convincing experiments were run with corn (Zea mays L.) which is not a host for H. virescens. Both α and β-DVT diols were sprayed on corn plants which were placed in cages. After exposure to moths, 1,798 eggs were found on treated plants while only 107 were present on controls (P<0.01) (Table III). Further evidence is presented from cage studies wherein 26 cultivars of tobacco, each containing different levels of α and β-DVT diols at the time of the test had egg deposits that were proportional to the amount of the compounds (Figure 3).

While the data base of these experiments is being expanded, it should be remembered that the isolation and purification of these metabolites is no easy accomplishment and that preparative costs are prohibitive. Field tests, if they are to be carried out on native plant populations without bias, must depend on the migratory habits of H. virescens and since the ovipositional attractants (α and β-DVT diols) are added as a bolus to the resistant plant their lifetime is probably short, with oxidation occurring in the first few hours (22).

The ovipositional stimulus is not limited to the α and β-DVT diols. Tobacco Introduction 1223 lacks the ability to produce these structures but, instead, has high concentrations of α and β-DVT ols (Figure 4). These have recently been investigated for

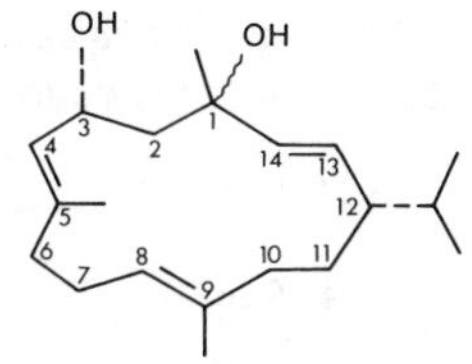

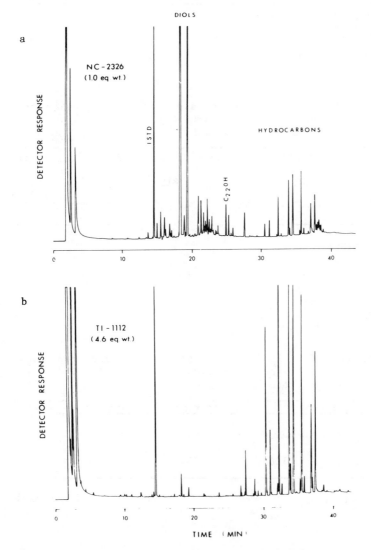

Figure 1. Structure of α and β- 4,8,13-duvatriene-1,3-diol

Figure 2. (a) Glass capillary gas chromatogram of the silylated cuticular compounds of tobacco NC 2326. (b) Glass capillary gas chromatogram of the silylated cuticular compounds of TI 1112.

Table I.   Data from 1980- 1983 cage trials with H. virescens
moths, on tobacco (Nicotiana tabacum L.)[a]

CULTIVAR	TREATMENT	TOTAL EGGS	PERCENT	F-TEST
TI-1112	NONE	6,610	23.6	
				P<0.01
NC 2326	NONE	21,424	76.4	

[a]113 REPLICATIONS

Table II. Data from 1982 cage trials with H. virescens moths,
showing oviposition stimulus with α and β-DVT diols, on
tobacco (Nicotiana tabacum L.)[a]

CULTIVAR	TREATMENT	TOTAL EGGS	PERCENT	F-TEST
TI-1112	ACETONE + WATER	811	27.8	
				P<0.05
TI-1112	α + β-DVT DIOLS	2,104	72.2	

[a]8 REPLICATIONS

Table III. Data from 1984 cage trials with H. virescens moths,
showing oviposition stimulus with α and β-DVT diols, on
non-host corn (Zea mays L.)[a]

PLANT	TREATMENT	TOTAL EGGS	PERCENT	F-TEST
CORN	ACETONE + WATER	107	5.6	
				P<0.01
CORN	α + β-DVT DIOLS	1,798	94.6	

[a]19 REPLICATIONS

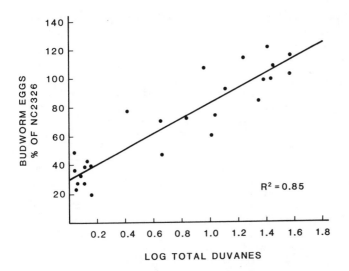

Figure 3. Concentration effect of α and β-DVT diols on deposition of eggs by <u>Heliothis virescens</u> on tobacco leaves in cage trials. Each dot represents a different tobacco cultivar. DVT levels are cultivar dependent.

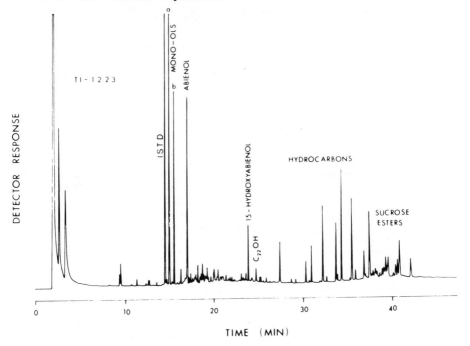

Figure 4. Glass capillary gas chromatogram of the silylated cuticular compounds of tobacco TI 1223.

their ovipositional properties. If TI 1112 is treated with the
α-and β-DVT ols, egg numbers are approximately doubled in cage
tests (Table IV) relative to untreated TI 1112 controls.

Plant and microbial responses with DVT diols and DVT ols. As we
have stated (1,2) the α-and β-DVT diols are potent inhibitors of
etiolated wheat coleoptiles. They do not appear to have any
antibiotic properties against the organisms E. coli, C. lunata,
and A. flavus. However, B. subtilis (gram positive) was inhibited
by α-DVT diol and zones of inhibition were 0, 15, and 15 mm with
50, 250 and 500 ug of the compound respectively but the β-DVT diol
did not induce inhibition. Also, Bacillus cereus (gram positive)
was inhibited by α-DVT and inhibition zones were 0, 11, and 14
mm with 50, 250, and 500 ug. Again, β-DVT diol did not inhibit
the organism. There is preliminary evidence to suggest that two
tobacco cultivars, Bel 61-10 and Chemical Mutant, both high
producers of α-and β-DVT diol, are highly resistant to tobacco blue
mold (Peronospora tabacina). We intend to further examine the
role of the DVT diols and ols with this organism.
    The α-and β-DVT ols are ovipositional stimulants for H.
virescens (Table IV) and also inhibit the growth of wheat
coleoptiles at 10^{-3} and 10^{-4} M, but because we have not yet
separated the α and β isomers the relative biological properties
of each isomer, cannot be established.

Antibiotic effects of the sucrose esters. Examination of the GC
profiles of the cuticular components of TI 165 reveals that small
quantities of α-and β-DVT ols and high quantities of α-and β-DVT
diols are present (Figure 5). From our earlier discussion it
would be predicted that TI 165 would be highly susceptible to
damage by H. virescens. It is, in fact, highly resistant.
Indeed, eggs of H. virescens are readily laid on TI 165, but
larvae die shortly after hatch (24). The reason for this oddity
is the production of a series of sucrose esters by TI 165, the
acids in association with the sucrose being acetic (always at the
C_6 position of the glucose moiety) and a series of C3 to C8 acids
(Table V) with the major acids being 2 methylbutyrate, and
3-methylvalerate (20). The ratio of acetic to the other acids is
always in the ratio of 1:3, respectively. Apart from the
antibiotic considerations, these substances are the key components
that impart the aroma and flavor characteristics to Turkish
tobacco (21, 22, 23).
    Several critical experiments have been made to determine if
newly hatched larvae will feed on the resistant TI 165 in a
similar fashion to that observed on susceptible NC 2326. Agar was
poured into 15 cm petri dishes to a depth of 5 mm to create a
humid atmosphere. A single leaf disk, each 25 mm in diameter,
was cut from each cultivar of TI 165 and NC 2326. These were
placed with the upper surface in the up position, on the agar, in
each plate. Five newly hatched larvae were introduced into each
dish and feeding scars were counted at 0.5, 1, 2, 4 and 24 hours
following introduction. There were no significant differences
between the numbers of feeding scars on either cultivar except at
2 hours when there were more feeding scars (P<0.01) on the
resistant TI 165. Therefore, the antibiosis effect is not due to

Table IV.   Data from 1982-1983 cage trials with **H. virescens** moths, showing oviposition stimulus ∝ and β-DVT ols, on tobacco (**Nicotiana tabacum** L.)[a]

CULTIVAR	TREATMENT	TOTAL EGGS	PERCENT	F-TEST
TI 1112	ACETONE + WATER	1,280	34.5	
TI 1112	∝ + β-DVT OLS	2,435	65.6	P<0.05

[a]20 REPLICATIONS

Table V.   Composition of fatty acids from a sucrose ester fraction of **N. tabacum** TI-165

ACIDS	MOLES/SUCROSE MOIETY
ACETIC	1.00
PROPIONIC	<0.01
ISOBUTYRIC	0.08
BUTYRIC	0.03
2-METHYLBUTYRIC	0.24
3-METHYLBUTYRIC	0.41
VALERIC	0.02
3-METHYLVALERIC	2.03
4-METHYLVALERIC	0.04
HEXANOIC	0.02
METHYLHEXANOIC	0.06
METHYLHEPTANOIC	0.01

MOLES ACETIC ACID: MOLES C_3 - C_8 ACIDS
                   1:3·0

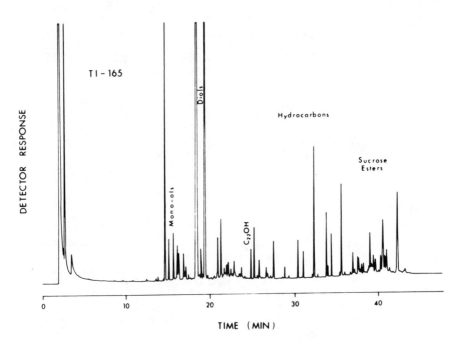

Figure 5.   Glass capillary gas chromatogram of the silylated
cuticular chemicals of tobacco TI 165.

a substance that inhibit feeding initiation by the larvae, which will begin to feed on resistant material if there is a limited food source available (25).

A further critical experiment utilized the cuticular surface compounds of TI 165 and NC 2326 which were partitioned against hexane:80% methanol-water. The composition of the polar phase was determined by capillary GC to be, predominantly, α and β -DVT diols and mixed sucrose esters. This material was added to the inside of 1 x 30 mm capillary tubes in 25 uL acetone (25). The exact composition of the materials contained in each capillary tube was, for TI 165, 53% α -DVT diol and 19% β-DVT diol, 5% oxy-duvatriene diols, 3% methylbutyric sucrose esters and 20% methylvaleric sucrose esters. The material in each capillary, for NC 2326, was 66% α -DVT diol, 23% β -DVT diol, 7% oxy-duvatriene diols, 4% methylbutyric sucrose esters and <1% methylvaleric sucrose esters. Each tube was assayed at the 'normal' concentrations of cultivar extracts from TI 165 and at twice and four times the normal rates. The normal levels for TI 165 are about twice that of NC 2326. The materials were uniformly deposited on the walls of the capillary tubes in a vacuum desiccator. An individual, newly hatched tobacco budworm larva was placed in each tube and survival rates were observed after 1, 2, 3 and 4 days. Controls consisted of acetone treated capillary tubes only, and capillary tubes alone. The effects of methanol-water extracts on larval longevity appeared to be approximately the same for NC 2326 and TI 165. After exposure for 2 hours there were visible signs of effects on larvae, especially at the two higher rates. Some larvae stuck to the exudate and became physically trapped. Other symptoms included quivering, twitching of the mouthparts and bubbling at the mouth. But while the results were approximately the same with extracts of both cultivars these experiments may not accurately reflect the physicochemical relationships at the leaf surface.

Artifical diet surfaces were treated with the methanol-water soluble extracts of TI 165 and NC 2326. Again, treatments were at the 1, 2 and 4-fold rates of that found on TI 165. These, too, were toxic to newly hatched budworm larvae and the LC_{50} for resistant (TI 165) and susceptible (NC 2326) cultivars was 188 and 231 ug of methanol-water soluble extract/cm^2 of diet surface (25). So it appears that both α and β-DVT diols in combination with the sucrose esters are important for antibiosis.

Plant and microbial responses with the sucrose esters. In addition to the above, we have conducted experiments on the effects of the sucrose esters on plants and microorganisms and we report these data for the first time. However, it should be reiterated that the sucrose esters were composed of acetic (C_2), 2-, and 3-methylbutyric (C5), and 3-methylvaleric (C6) acids and only the glucose molecule was esterified, the fructose portion remained in the free hydroxyl state. Also, the sucrose esters were extremely difficult to separate so that mixtures were initially obtained and tested. We shall see, in the wheat coleoptiles bioassay, that as the number of 3-methylvalerate residues decreases in the glucose moiety, so the level of biological activity decreases at the 10^{-4} M level, though in all cases there is 100% inhibition at 10^{-3} M (Figure 7).

In Figure 6(a) the sucrose mixture consists of glucose mixed with 92% one C_2 and three C_6 plus 8% of one C_2 and one C_5 plus two C_6; at 10^{-4} M, coleoptiles are inhibited 90% (Figure 7(a)). Figure 6(b) indicates a mixture of 69% one C_2 and three C_6 plus 31% one C_2 and one C_5 plus two C_6; at 10^{-4} M, coleoptiles are inhibited 83% (Figure 7(b)). Figure 6(c) shows that a mixture of 8% one C_2 and three C_6 and 85% one C_2 plus one C_5 plus two C_6; at 10^{-4} M coleoptiles are inhibited 60% (Figure 7(c)). Figure 6(d) states a mixture consisting of 20% of C_2 and one C_5 and two C_6 plus 77% one C_2 plus two C_5 and one C_6, plus 3% one C_2 and three C_5; at 10^{-4} M, coleoptiles are inhibited 43% (Figure 7(d)). Figure 6(e) indicates a mixture of 13% one C_2 plus one C_5 plus two C_6 and 81% one C_2, two C_5 and one C_6, plus 6% one C_2 and three C_5; coleoptiles were inhibited 13.5% at 10^{-4} M (Figure (7e)). In all cases inhibitions were significant (P<0.01) and percent inhibitions were relative to controls. The exact positions of the C_5 or C_6 acids has not been identified.

The pure sucrose ester (98%[+] purity) containing 3 residues of 3-methylvalerate is active against wheat coleoptile at 10^{-3} and 10^{-4} M and significantly inhibits them 100 and 81%, respectively (Figure 7e). But why is the 3-methylvalerate sucrose ester inhibitory to wheat coleoptiles? The phosphate-citrate solution in which the coleoptiles are incubated for 18 hours already contains 2% sucrose. Additional sucrose from the sucrose esters, assuming that de-esterification occurred, would not be enough to cause exosmosis from the plant cells thereby inducing apparent shortening of the segments. Therefore, it would seem that 3-methylvaleric acid is the inhibitory substance. Accordingly, 3-methylvaleric acid was included in the coleoptile bioassay at 10^{-3} to 10^{-6} M, but there was no inhibition. The odor of 3-methylvaleric acid was preceived in the 10^{-3} M solutions. The bioassay was also run using 3-methylvaleric acid in distilled water only to preclude the possibility of sucrose-valerate interactions, but no inhibitory response was obtained. And because of the possibility of insolubility, potassium 3-methylvalerate was prepared and coleoptile sections were assayed in distilled water in the presence of this salt. There was no inhibition. We obtained a sample of the glucose 3-methylvalerate ester (Figure 9,10) and assayed it against wheat coleoptiles from 10^{-3} to 10^{-6} M, but growth inhibition was not induced by this substance. Because the fructose 3-methylvalerate esters were not found in tobacco sucrose esters, they were not tested. Thus, it would appear the combination of two molecules (sucrose and 3-methylvalerate), each inactive as growth inhibiting substances at the concentrations tested in the bioassay, is necessary for induction of biological activity. Certainly, the question may be asked as to why the combination has to occur with sucrose. One possibility is that the sucrose esters interfere with the sucrose synthetase enzyme system, but this remains to be demonstrated.

None of the sucrose ester mixtures were active against the fungi C. lunata and A. flavus, neither were they active against the gram negative E. coli and, in later experiments, E. cloacae and C. freundii. However, they were active against B. subtilis and, as with the wheat coleoptile bioassays, the activity appeared to be related to the number of 3-methylvalerate residues on the

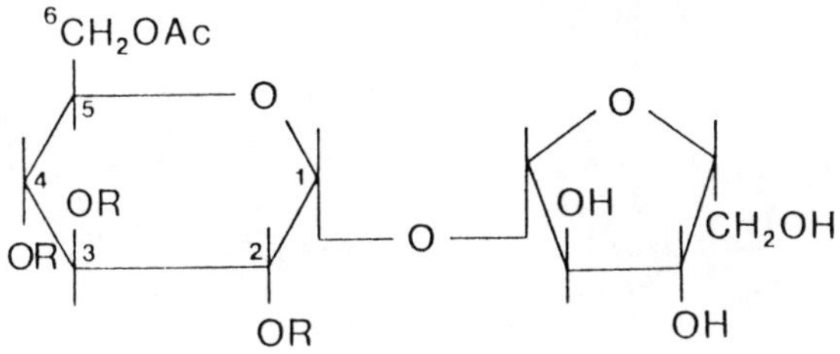

$^{6}CH_2OAc$

Acid substitution (#carbons)

(a) 92% one C2 & three C6
     8% one C2 & one C5 & two C6´

(b) 69% one C2 three C6
     31% one C2 & one C5 & two C6

(c) 8% one C2 & three C6
     85% one C2 & one C5 & two C6
     7% one C2 & two C5 & one C6

(d) 20% one C2 & one C5 & two C6
     77% one C2 & two C5 & one C6
     3% one C2 & three C5

(e) 13% one C2 & one C5 & two C6
     81% one C2 & two C5 & one C6
     6% one C2 & three C5

Figure 6. Sucrose ester mixtures isolated from the cuticular components of tobacco TI 165 (Refer to Table V for details).

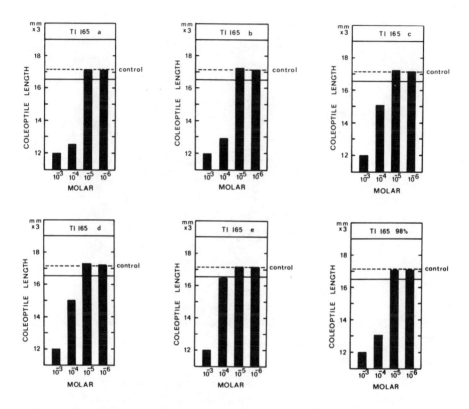

Figure 7. Effects of sucrose ester mixtures isolated from the cuticular components of tobacco TI 165 on the growth of etiolated wheat (T. aestivum L., cv Wakeland) coleoptiles. Significant inhibition (P<0.01): below solid line. Control: broken line. (Refer to Figure 6 for details).

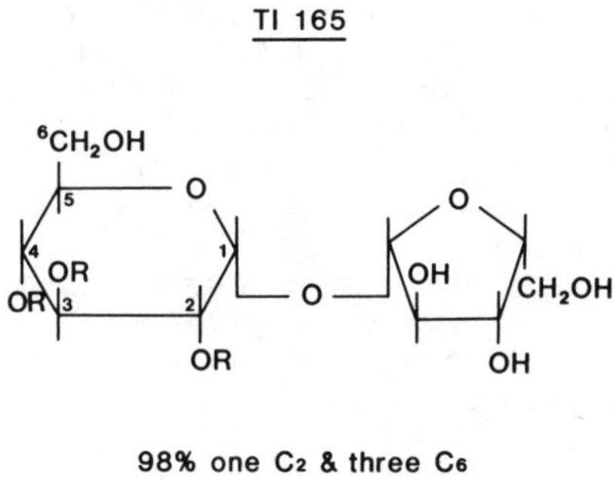

TI 165

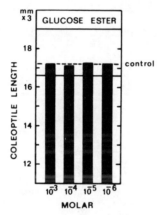

98% one C₂ & three C₆

Figure 8. Structure of the pure sucrose ester from tobacco TI 165. The C_6 position has an acetate; R=3-methylvalerate.

Figure 9. Effects of pure glucose 3-methylvalerate on the growth of etiolated wheat (**T. aestivum** L., cv Wakeland) coleoptiles. No significant inhibition.

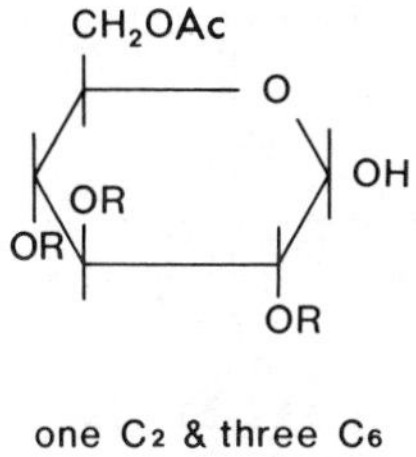

one C₂ & three C₆

Figure 10. Structure of synthetic glucose 3-methylvalerate.

glucose portion of the sucrose molecule (Table VI). Though the
concentrations used were slightly different in each case, note
that TI 165(b) requires almost twice as much mixed sucrose ester
as TI 165(a) to induce inhibition (Table VI), and that TI 165(e)
exhibits no activity. In addition to inhibiting B. subtilis, the
98% pure sucrose ester inhibited gram positive B. cereus (9, 11,
and 11 mm zones of inhibition at 50, 250 and 500 ug doses,
respectively) and the gram positive Mycobacterium thermosphactum
(0, 8, and 10 mm zones of inhibition at 50, 250 and 500 ug doses,
respectively). The latter organism is a close relative of
Mycobacterium tuberculosis, the agent responsible for
tuberculosis, a disease still prevalent throughout the world even
in industrial nations. The potential medicinal properties of
these compounds warrants investigation.

Table VI. Antibiotic effect of sucrose esters on Bacillus
          subtilis in petri dish bioassays[a].

COMPOUND[b]	RATE/DISK (uG)	INHIBITION ZONE (MM DIAMETER)
TI 165 (a)	49	0
	245	10
	490	12
TI 165 (b)	80	0
	400	12
	800	13
TI 165 (c)	53	0
	265	11
	530	11
TI 165 (d)	58	0
	290	0
	580	11
TI 165 (e)	64	0
	320	0
	640	0
3-METHYL VALERIC ACID	50	0
	250	0
	500	0
98% PURE SUCROSE ESTER	50	10
	250	11
	500	14

[a] Duplicate series
[b] See Figure 6 for Legend

    Perhaps the most striking aspect of the studies with
microorganisms is the selective activity of the sucrose esters
against the gram positive bacteria and the apparent inactivity
against gram negative bacteria and fungi. Limited amounts of the

sucrose esters precluded larger assays and observations on greater numbers of gram positive and gram negative bacteria. Even in the bacterial bioassays, 3-methylvaleric acid, potassium 3-methylvalerate, and glucose 3-methylvalerate, were inactive. Thus, the unique combination of the acid (3-methylvalerate) esterified with sucrose is important to the inhibitory properties of the molecule. It is interesting to note that most of the bacterial pathogens of N. tabacum are gram negative and the question arises as to the role of trichome exudates in controlling plant pathogens. Again, we would emphasize that the sucrose esters do not have high specific activity in our plant and bacterial bioassays, but under natural conditions these compounds are produced in a continuum by the trichomes, whereas materials are applied in our bioassays as a bolus. Also, initial studies indicate that the sucrose esters are far more stable than their associated cuticular terpenes.

The leaf surface of tobacco produces many interesting and varied compounds that possess interesting biological properties. We are presently studying other sucrose esters from tobacco plants that are inhibitory to etiolated wheat coleoptiles and which are more active than the 3-methylvalerate sucrose esters against microorganisms. They will be the topic of future reports.

Literature Cited

1. Cutler, H.G. Science 1970, 170, 856.
2. Cutler, H.G.; Cole, R.J. Plant Cell Physiol. 1974, 15, 19.
3. Springer, J.P.; Clardy, J.; Cox, R.H.; Cutler, H.G.; Cole, R.J. Tetrahedron Lett. 1975, 32, 2734.
4. Roberts, D.L.; Rowland, R.L. J. Org. Chem. 1962, 27, 3989.
5. Addicott, F.T. In "Abscisic Acid"; Praeger Pubs. (Holt, Reinhart and Winston): New York 1983.
6. Cutler, H.G.; Reid, W.W.; Deletang, J. Plant Cell Physiol. 1977, 18, 711.
7. Perkins, D.L.; Cieresko, L.S. Proc. Okla. Acad. Sci. 1974, 54, 34.
8. Severson, R.F.; Arrendale, R.F.; Chortyk, O.T.; Johnson, D.M.; Gwynn, G.R.; Chaplin, J.F.; Stephenson, M.G. J. Agric. Food Chem. 1984, 32, 566.
9. Severson, R.F.; Arrendale, R.F.; Chortyk, O.T.; Johnson, A.W. 32nd Southern Regional American Chemical Society Meeting, Lexington, KY, 1981.
10. Severson, R.F.; Chortyk, O.T.; McDuffie, K.L.; Johnson, A.W.; Gwynn, G.R.; Chaplin, J.F. 35th Tobacco Chemist's Research Conference, Winston-Salem, NC, 1981.
11. Severson, R.F.; McDuffie, K.L.; Jackson, D.M.; Johnson, A.W.; Stephenson, M.G.; Herzog, G.A. J. Ga. Acad. Sci. 1982, 40, 15.
12. Nitsch, J.P.; Nitsch, C. Plant Physiol. 1956, 31, 94.
13. Cutler, H.G. In "Bioregulators: Chemistry and Uses", Eds. R.L. Ory, F.R. Rittig. Chapter 14. ACS Symposium Series 257, Washington, DC. 1984.
14. Kurtz, T.E.; Link, R.F.; Tukey, J.W.; Wallace, D.L. Technonmetrics 1965, 7, 95.
15. Cutler, H.G. 11th Proc. Plant Growth Regulator Soc. America 1984, 1.

16. Elsey, K.D.; Chaplin, J.F. J. Econ. Entomol. 1978, 71, 723.
17. Johnson, A.W.; Chaplin, J.F. Tob. Sci. 1982, 26, 157.
18. Jackson, D.M.; Severson, R.F.; Johnson, A.W.; Chaplin, J.F.; Stephenson, M.G. Environmental Entomol. 1984, 13, 1023.
19. Crosby, D.G.; Vlitos, A.J. In "Plant Growth Regulation," Iowa State University Press, Iowa, 1961, p. 57.
20. Severson, R.F.; Arrendale, R.F.; Chortyk, O.T.; Green, C.R.; Thome, F.A.; Stewart, J.L.; Johnson, A.W. J. Agric. Food Chem. 1985, (In Press).
21. Kallianos, A.G. Recent Advances Tob. Sci. 1976, 2, 61.
22. Chuman, T.; Noguchi, M. Agric. Biol. Chem. 1977, 41, 1020.
23. Matsushima, S.; Ishiguro, S.; Sugawara, S. Beitr. Tabakforsch. 1979, 10, 31.
24. Johnson, A.W.; Severson, R.F. J. Agric. Entomol. 1984. 1, 23.
25. Johnson, A.W.; Severson, R.F. Southeastern Branch, Entomogical Soc. of America, Greenville, SC, January 28-31, 1985.

RECEIVED October 23, 1985

Photoactivated Biocides from Higher Plants

Kelsey R. Downum

Department of Biological Sciences, Florida International University, Miami, FL 33199

Plants synthesize chemicals which possess a wide range
of biological activities. Natural products that
require light for expression of their toxic biological
consequences are among the most interesting of these
bioactivities. Such "photosensitizers" or "phototoxins"
are structurally variable, are produced by widely
divergent flowering plant families and mediate broad-
spectrum toxic reactions. Acetophenones, alkaloids,
furocoumarins, furochromones, polyines as well as one
important lignan have been added to a growing list of
phototoxic metabolites isolated from plant sources.
The chemistry, distribution and biocidal action of
these toxic allelochemicals is discussed.

Plants produce an array of chemicals that can be excited by sunlight.
The best known of these natural products include chlorophylls,
carotenoids and phytochrome. Chlorophylls and carotenoids
participate in photosynthesis, the process evolved by plants to
convert light into the chemical energy essential for biosynthetic
processes. Phytochrome, on the other hand, mediates a variety of
growth and developmental responses such as seed germination, growth
regulation and floral initiation. In addition to light-activated
"pigments", many plants also produce "photosensitizers" or
"phototoxins". These metabolites are unique in that they become
toxic to other organisms in the presence of sunlight (specifically
the UVA region of sunlight; 320-400 nm). The biocidal potential of
such phototoxic phytochemicals has been demonstrated against microorganisms (including important bacterial and fungal pathogens of
plants), nematodes, herbivorous and non-herbivorous insects as well
as numerous other organisms (1-3). Such broad-spectrum bioactivity
suggests that plant photosensitizers may function as "solar-powered"
defensive agents which discourage insect herbivory and inhibit
infection by pathogenic organisms in nature.

0097-6156/86/0296-0197$06.00/0

Chemistry of Photosensitizers

Plant metabolites that are capable of becoming toxic in sunlight or
UVA are produced by a wide variety of biochemical pathways and thus
are a structurally diverse group of natural products. Both linear
and cyclic photosensitizers are known from plants. Linear
phototoxins are generally derived from fatty acid precursors and
typically possess conjugated double and triple bond systems. The
majority of cyclic photosensitizers, on the other hand, are bi- and
tricyclic aromatic molecules that may contain nitrogen, oxygen or
sulfur as heterocyclic elements.

Most phototoxic phytochemicals are cyclic. Various alkaloids,
acetophenones, extended quinones, furochromones and furocoumarins
belong to this class. Thiophenic and ring-stabilized acetylenic
polyines as well as the lignan nordihydroguaiaretic acid (NDGA)
belong to a second group of photosensitizers that have both linear
and cyclic moieties while a third type consists of straight-chain
molecules only. Examples of photosensitizers from these categories
are given in Figure 1.

Distribution of Photosensitizers in Plants

Chemicals that are potentially capable of phototoxic action have
been isolated from more than 30 flowering plant families. Their
occurrence among important monocot and dicot families is shown in
Table I. Most of the taxa represented in this table synthesize
several types of photosensitizers. Members of the Asteraceae
(sunflower family) and the Rutaceae (citrus family) for example
synthesize the widest range of phototoxic compounds. Other families
(e.g., Hypericaceae, Liliaceae, Moraceae and Orchidaceae) either
lack or fail to express such biosynthetic diversity. Plants from
these latter groups contain phototoxins derived from a single
metabolic pathway.

Acetylenes, beta-carboline alkaloids, furocoumarins and lignans
occur widely among the families listed in Table I. Acetophenones
(benzofurans and benzopyrans) and extended quinones have a much more
limited distribution while furochromones, furoquinoline alkaloids
and thiophenes are apparently restricted to single families. It
should be mentioned that although acetylenic polyines are
phytochemical components of plants from numerous families (4), only
those derivatives produced by members of the Asteraceae have been
shown to be phototoxic (1).

Lignans are included in Table I because they represent an
important class of potential photosensitizers. We recently isolated
and identified the first phototoxic lignan nordihydroguaiaretic acid
from the leaf resin of the creosote bush Larrea tridentata
(Zygophyllaceae) (5). A total of more than 200 different lignans
from ca. 60 plant families (6) are known and represent a significant
pool of potentially active phytochemicals. Further investigation
is needed to establish whether this biological activity is unique
to NDGA or representative of the action of other lignans.

The ecotypic distribution of major phototoxin-producing
families is shown in Table II. Plants that synthesize these
biocidal components are widely distributed in nature (after ref. 12),

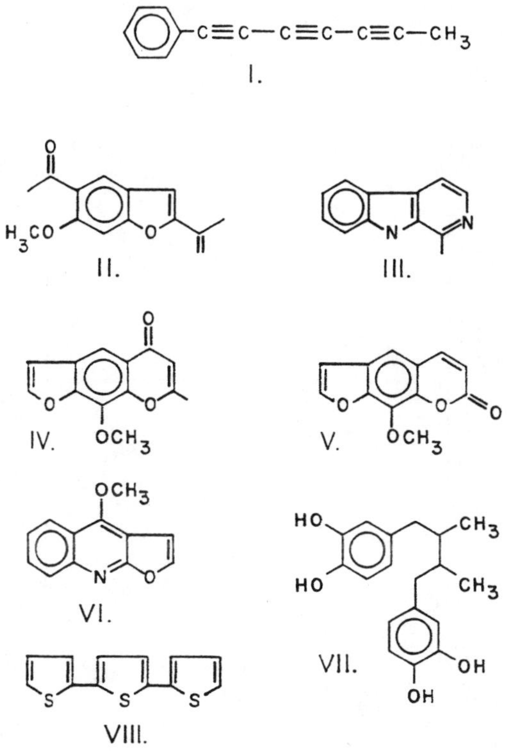

Figure 1.   Structures of representative phototoxic plant products
from various phytochemical classes.   (I) 1-phenylhepta-1,2,3-
triyne (acetylenic polyine); (II) 6-methoxyeuparin (benzofuran);
(III) harmane (beta-carboline alkaloid); (IV) khellin (furo-
chromone); (V) 8-methoxypsoralen (furocoumarin); (VI) dictamnine
(furoquinoline alkaloid); (VII) nordihydroguaiaretic acid
(lignan) and (VIII) alpha-terthienyl (thiophene).

Table I. Distribution of Phototoxic Phytochemicals

Plant Family	I	II	III	IV	V	VI	VII	VIII	IX	X	References
Apiaceae	+					+			+		4,6-7
Araliaceae	+								+		4,6
Asteraceae	+	+				+			+	+	4,6-7,9-11
Cyperaceae		+									8-9
Euphorbiaceae	+							+			4,12
Fabaceae	+		+			+					4,7,9
Hypericaceae				+							13,14
Liliaceae		+									8
Moraceae						+					7
Orchidaceae						+					7
Polygonaceae			+	+					+		6,9,13
Rubiaceae			+					+			9,12
Rutaceae	+	+	+		+	+	+	+	+		1,4,6-9,12
Solanaceae	+		+			+			+		4,6-7,9
Zygophyllaceae			+						+		6,9

I. Acetylenes; II. Acetophenones; III. Beta-Carboline Alkaloids; IV. Extended Quinones; V. Furochromones; VI. Furocoumarins; VII. Furoquinoline Alkaloids; VIII. Isoquinoline Alkaloids; IX. Lignans; X. Thiophenes.

but seem to occur most often in tropical/subtropical climates. Such familial distributions are quite interesting and suggest that plants with endogenous phototoxins may be most successful in environments where they are (or could be) exposed to intense solar irradiation throughout much of the year.

Table II. Distribution of Phototoxin-Containing Plant Families

Plant Families	Main Distribution
Dicotyledonous	
Apiaceae (carrot & parsnips family)	Temperate Uplands
Araliaceae (ivy & ginseng family)	Tropics
Asteraceae (sunflower family)	World-Wide
Euphorbiaceae (spurge family)	Tropics
Hypericaceae	Tropics to Temperate
Moraceae (fig family)	Tropics/Subtropics
Polygonaceae (buckwheat family)	Temperate
Rubiaceae (coffee family)	Tropics/Subtropics
Rutaceae (citrus family)	Tropics to Temperate
Solanaceae (potatoe family)	Tropics to Temperate
Zygophyllaceae	Tropics/Subtropics
Monocotyledonous	
Cyperaceae (reeds & sedges)	Temperate
Liliaceae (lily family)	World-Wide
Orchidaceae (orchid family)	Tropics

Biocidal Action of Photosensitizers

The toxic mechanisms of photosensitization have been reviewed recently (1-3) and will not be dealt with here. The reader is referred to these references for detailed discussions of the cellular targets and molecular mechanisms of phototoxicity mediated by various natural products.

The broad-spectrum biocidal action induced by phototoxic allelochemicals has been demonstrated toward a range of organisms including bacteria, fungi, nematodes, insects as well as non-phototoxin-containing plants. Table III summarizes the organisms susceptible to photosensitization by alpha-terthienyl, a phototoxin characteristic of many Asteraceae species (4, 10-11). The toxicity of this thiophenic polyine toward plant pathogens (e.g., Agrobacterium, Alternaria, Cladosporium, Colletotrichum, Fusarium, Pythium, Rhizoctonia), nematodes and herbivorous insects (e.g., Euxoa, Heliothus, Manduca, Ostrinia) suggests that it may function in nature to discourage not only herbivory, but also infection by potential pathogens. Many natural products react with similar non-specificity in UVA (34-44). The function(s) of these allelochemicals remains to be demonstrated, but deterrence of deleterious organisms seems to be a possibility worthy of further investigation.

The susceptibility of particular bacterial or fungal pathogens

to phototoxic phytochemicals is based exclusively on in vitro
studies. Efforts to evaluate the involvement of these photosensitizers in plant resistance to disease organisms have not yet been
conducted. The absence of a well defined host-pathogen system for
study is one of the principal reasons for this void in our
understanding.

Table III. UVA-Mediated Lethal Activity of Alpha-Terthienyl
Toward Various Organisms

Biocidal Activity	Susceptible Organisms	References
Bactericidal	Agrobacterium, Bacillus, E. coli Pseudomonas, Staphlococcus	16-21
Fungicidal	Alternaria, Aspergillus, Candida, Cladosporium, Colletotrichum, Fusarium, Pythium, Rhizoctonia, Rhizopus, Saccharomyces, Saprolegnia	17-18,22-23
Nematicidal	Aphelenchus, Caenorhabditis, Ditylenchus, Meloidogyne, Pratylenchus	24-25
Insecticidal	Aedes, Euxoa, Heliothus, Manduca, Ostrinia, Simulium, Spodoptera	26-31
Allelopathic	Asclepias, Chenopodium, Lactuca, Phleum, Trifolium	32-33

    We have recently begun to look for new plant species that
contain endogenous photosensitizers which might be useful in disease
resistance studies. Preliminary efforts have centered on screening
previously untested plants for UVA-induced antibiotic activity.
Agriculturally important species from phototoxin-containing families
were given the highest priority for testing as details concerning
their pathology and phytochemistry would most likely be available.
Citrus and several closely related genera in the Rutaceae were among
the most active of the plants selected for examination. Methanolic
leaf extracts from eight genera were applied to sterile filter paper
discs and bioassayed (in UVA and dark) on agar plates using standard
microbial techniques (described in ref. 45). The results of these
bioassays are given in Table IV. Extracts from six genera were
phototoxic to E. coli and to the yeast Saccharomyces cerevisiae
while extracts of Fortunella and Glycomis species were not.
Antibiotic activity in the absence of UVA was not noticeable with
any of the leaf extracts.
    These findings are of particular note because they are the first
report of phototoxic action in the genera Afraegle, Atalantia,
Citrus, Microcitrus, Severinia and Swinglea although photosensitizers
have been demonstrated in other genera of the family (7). Various
furocoumarin derivatives which have been isolated from Citrus
species (46-48) are undoubtedly responsible for much of the
bioactivity of the species assayed, however the role that other

photosensitizers might play in this lethal action has not been established.

Table IV.   UVA-Induced Antibiotic Action of Leaf Extracts of Citrus
and Related Genera of the Rutaceae

Plant Species	E. coli Dark	E. coli UVA	S. cerevisiae Dark	S. cerevisiae UVA
Afraegle paniculata (Swing.) Engler	−	+++	−	+++
Atalantia monophylla DC.	−	++	−	+++
Citrus depressa Hayata (mandarin)	−	++	−	++
C. grandis (L.) Osbeck (pummelo)	−	+	−	+
C. limetta Risso (limetta)	−	++	−	++
C. limettoides Tan. (sweet lime)	−	++	−	+++
C. macrophylla Wester (alemow)	−	++	−	++
C. medica L. (citron)	−	++	−	++
C. miaray Wester	−	+	−	+
C. sinensis (L.) Osbeck (sweet orange)	−	+	−	+
C. tiawanica	−	+	−	+
C. ujukitsu Hort. ex Tan.	−	+	−	+
Glycomis pentaphylla (Retz.) Correa	−	−	−	−
Fortunella sp. (kumquat)	−	−	−	−
Microcitrus australasica (F. Muell.) Swing. (Australian finger-lime)	−	++	−	++
Severinia buxifolia (Poir.) Tenore (Chinese box-orange)	−	++	−	++
Swinglea glutinosa (Blanco) Merr. (tabog or swinglea)	−	++	−	++

No antibiotic activity (−); Inhibitory zones below 14 mm (+),
between 15-20 mm (++) and larger than 21 mm (+++).

Many Citrus pathogens are known including species of Alternaria,
Cercospora, Fusarium, Mycosphaerella, Phytophthora, Pseudomonas and
Xanthomonas (citrus canker). The susceptibility of several of
these organisms to endogenous Citrus photosensitizers is currently
under investigation in my laboratory. Susceptible pathogens, based
on in vitro bioassays, will be used to evaluate the importance of
phototoxins in the resistance of Citrus species to infection by
pathogenic organisms.

Conclusion

The number and structural diversity of phototoxic phytochemicals has
grown tremendously over the last ten years. Our knowledge and
understanding of the biological activity of these natural products
will undoubtedly continue to expand as new structures are elucidated
and new plant families are examined. In addition to exploratory
studies, it is important that we begin to turn our attention toward
the significance or function of photosensitizers. Do plants which

contain these chemicals have selective advantages over species which
lack them? Do increased levels of a particular photosensitizer
increase a plants resistance to pathogens, nematodes or insect
herbivory? If so, are their autotoxic effects expressed in tissues
containing high levels of phototoxins? Answers to these and other
questions are necessary in order to clarify the role(s) of phototoxic
natural products.

Acknowledgments

I would like to thank Drs. Eloy Rodriguez and G.H.N. Towers for
critical reading of this manuscript, Dr. C. Campbell (Tropical
Research and Education Center-IFAS, University of Florida, Homestead,
FL) for access to the various Citrus species and related Rutaceae
genera used in this study and J.A. Downum for field and technical
assistance. The support of NSF (PCM 8209100) and NIH (AI18398) to
E. Rodriguez and the FIU Foundation (571204900) is gratefully
acknowledged.

Literature Cited

1. Towers, G.H.N. Can. J. Bot. 1984, 62, 2900-11.
2. Downum, K.R.; Rodriguez, E. J. Chem. Ecol. In press.
3. Knox, J.P.; Dodge, A.D. Phytochem. 1985, 24, 889-96.
4. Bohlmann, F.; Burkhardt, T.; Zdero, C. "Naturally Occurring
   Acetylenes"; Academic: London, 1973.
5. Downum, K.R.; Dole, J.; Rodriguez, E. Phytochem. In review.
6. MacRae, W.D.; Towers, G.H.N. Phytochem. 1984, 23, 1207-21.
7. Murray, R.D.H.; Mendez, J.; Brown, S.A. "The Natural Coumarins:
   Occurrence, Chemistry and Biochemistry"; Wiley: New York, 1982.
8. Proksch, P.; Rodriguez, E. Phytochem. 1983, 22, 2335-48.
9. Allen, J.R.F.; Holmstedt, B.R. Phytochem. 1980, 19, 1573-82.
10. Downum, K.R.; Towers, G.H.N. J. Nat. Prod. 1983, 44, 98-103.
11. Downum, K.R.; Keil, D.J.; Rodriguez, E. Biochem. Syst. Ecol.
    In press.
12. Raffauf, R.F. In "A Handbook of Alkaloids and Alkaloid-Contain-
    ing Plants"; Wiley: New York, 1970.
13. Thompson, R.H. "Naturally Occurring Quinones"; Academic: London,
    1971.
14. Arnason, T.; Towers, G.H.N.; Philogene, B.J.R.; Lambert, J.D.H.
    In "Plant Resistance to Insects"; Hedin, P.A., Ed.; ACS
    SYMPOSIUM SERIES No. 208, American Chemical Society: Washington,
    D.C., 1983, pp. 139-51.
15. Heywood, V.H. In "Flowering Plants of the World"; Mayflower: New
    York, 1978.
16. Camm, E.L.; Towers, G.H.N.; Mitchell, J.C. Phytochem. 1975, 14,
    2007-11.
17. Chan, G.F.Q.; Towers, G.H.N.; Mitchell, J.C. Phytochem. 1975.
    14, 2295-6.
18. Arnason, T.; Chan, G.F.Q.; Wat, C.-K.; Downum, K.; Towers, G.H.N.
    Photochem. Photobiol. 1981, 33, 821-4.
19. Downum, K.R.; Hancock, R.E.W.; Towers, G.H.N. Photochem.
    Photobiol. 1982, 36, 517-23.

20. Downum, K.R.; Hancock, R.E.W.; Towers, G.H.N. Photobiochem. Photobiophysics 1983, 6, 145-152.
21. Norton, R.A. Personal communication.
22. DiCosmo, F.; Towers, G.H.N.; Lam, J. Pest. Sci. 1982, 13, 589-94.
23. Kourany, E.; Arnason, J.T. Personal communication.
24. Gommers, F.J.; Geerlings, J.W. Nematologica 1973, 19, 389-93.
25. Gommers, F.J.; Bakker, J.; Wynberg, H. Photochem. Photobiol. 1982, 35, 615-19.
26. Wat, C.-K.; Prasad, S.K.; Graham, E.A.; Partington, S.; Arnason, T.; Towers, G.H.N. Biochem. Syst. Ecol. 1981, 59-62.
27. Arnason, T.; Swain, T.; Wat, C.-K.; Graham, E.A.; Partington, S.; Towers, G.H.N. Biochem. Syst. Ecol. 1981, 9, 63-8.
28. Kagan, J.; Chan, G. Experientia 1983, 39, 402-3.
29. Kagan, J.; Chan, G.; Dhawan, S.N.; Arora, S.K.; Prokash, I. J. Nat. Prod. 1983, 46, 646-50.
30. Downum, K.R.; Rosenthal, G.A.; Towers, G.H.N. Pest. Biochem. Physiol. 1984, 22, 104-9.
31. Champagne, D.E.; Arnason, J.T.; Philogene, B.J.R.; Morand, P.; Lam, J. J. Chem. Ecol. In Press.
32. Campbell, G.; Lambert, J.D.H.; Arnason, T.; Towers, G.H.N. J. Chem. Ecol. 1982, 8, 961-72.
33. Downum, K.R.; Quiroz, A.M.; Rodriguez, E.; Towers, G.H.N. Unpublished results.
34. Berenbaum, M. Science 1978, 201, 532-4.
35. Berenbaum, M. Ecol. Entomol. 1981, 6, 345-51.
36. Friedman, J.; Rushkin; Walker, G.R. J. Chem. Ecol. 1982, 8, 55-65.
37. Fujita, H.; Ishii, N.; Suzuki, K. Photochem. Photobiol. 1984, 39, 831-4.
38. McKenna, D.J.; Towers, G.H.N. Phytochem. 1981, 20, 1001-4.
39. Muckensturm, B.; DuPlay, D.; Robert, P.C.; Simonis, M.T.; Kienlen, J.-C. Biochem. Syst. Ecol. 1981, 9, 289-92.
40. Oginsky, E.L.; Green, G.S.; Griffith, D.G.; Fowlks, W.L. J. Bacteriol. 1959, 78, 821-33.
41. Shimomura, H.; Sashida, Y.; Nakata, H.; Kawasaki, J.; Ito, Y. Phytochem. 1982, 21, 2213-5.
42. Towers, G.H.N.; Graham, E.A.; Spenser, I.D.; Abramowski, Z. Planta Medica: J. Med. Plant. Res. 1981, 41, 136-42.
43. Abeysekera, B.F.; Abramowski, Z.; Towers, G.H.N. Photochem. Photobiol. 1983, 38, 311-5.
44. Towers, G.H.N.; Abramowski, Z. J. Nat. Prod. 1983, 46, 576-81.
45. Downum, K.R.; Villegas, S.; Keil, D.J.; Rodriguez, E. Biochem. Syst. Ecol. In press.
46. Kefford, J.F.; Chandler, B.V. In "The Chemical Constituents of Citrus Fruits"; Academic: New York, 1970, pp. 106-11.
47. Dreyer, D.L.; Huey, P.F. Phytochem. 1973, 12, 3011-13.
48. Fisher, J.F.; Trama, L.A. Agric. Food Chem. 1979, 27, 1334-37.

RECEIVED August 9, 1985

17

Insect Ecdysis Inhibitors

Isao Kubo and James Klocke[1]

Division of Entomology and Parasitology, College of Natural Resources, University of California, Berkeley, CA 94720

A requisite for developmental growth in insects is molting, the entire process of the shedding of the old cuticle. The process of molting is initiated when the insect molting hormone, ecdysterone or 20-hydroxyecdysone (Figure 1), stimulates the epidermis to retract from the cuticle. This retraction, termed apolysis, is immediately followed by mitotic division of the epidermal cells and their subsequent secretion of both a protective procuticle and a gel-like molting fluid. Following activation of the molting fluid, enzymatic digestion of the old cuticle for resorption and reuse results in a thin (undigested) remnant of the old exocuticle, termed exuvia, which is subsequently split and cast off. The casting off of the exuvia, termed ecdysis, is accomplished by hydrostatic pressures brought about by the swallowing of air or water and the performance of muscular activities (1).

When new cuticle is synthesized, it is soft and flexible so that the hydrostatic pressures unfold and expand the new cuticle thereby increasing its surface area and concomitantly casting off the old cuticle. After ecdysis, expansion of the new cuticle is brought to an end by the onset of sclerotization (tanning) (2), which involves the cross-linking of cuticular protein with ortho-quinone. The source of the ortho-quinone is tyrosine, whose mobilization is controlled by ecdysterone, as well as by a peptide hormone called bursicon.

The major events of the molting process are apolysis, cuticular synthesis, ecdysis, and sclerotization (3, 4). The complexity of the sequence of physiological and developmental events occurring in the molting process, and the high degree of hormonal coordination with which the entire process must proceed, render the insect which must molt particularly vulnerable to exogenously applied chemicals. Molts are recurrent crises in the lives of arthropods (5). Chemicals which can trigger any of the major events of the molting process out of their normal sequence may be useful as insect control agents both because of their adverse effects on molting and also because of their possible

[1]Current address: NPI-Plant Resources Institute, 417 Wakara Way, Salt Lake City, UT 84108.

0097-6156/86/0296-0206$06.00/0
© 1986 American Chemical Society

specificity for organisms which molt (e.g., insects). The present
manuscript will discuss those synthetic and natural chemicals
which inhibit or prevent ecdysis and, as such, either are presently
available as or are candidates for insect control agents.

The actual shedding of the old cuticle during molting is
termed ecdysis. Ecdysis is triggered by a neurosecretory peptide,
eclosion hormone, which acts on the central nervous system to
elicit the ecdysial motor programmes (6). The release of eclosion
hormone is in turn regulated by the titre of the molting hormone,
ecdysterone (7). The inhibition of ecdysis, which can occur
through a disruption of the normal titres of ecdysterone or
eclosion hormone or through other factors in the complex pathway,
is easily observed as a gross morphological event in which the
"old" cuticle remains enveloping the insect in the pharate
condition. The abnormal pharate condition prevents feeding and
locomotion, and eventually results in the death of the affected
insect. The inhibited ecdysis of a <u>Bombyx</u> <u>mori</u> (silkworm)
larva is shown in Figure 2. The larva underwent normal apolysis,
but failed to complete ecdysis and could not remove its head
capsule or trunk exuvia and eventually died.

Until recently, most of the compounds which were found to
interfere with ecdysis were steroids, either identical to or
structurally similar to ecdysterone. Ecdysterone, and many analogs
of it, have been isolated from both animal (zooecdysones) (8) and
plant (phytoecdysones) (9) sources. Some of these compounds have
been found to inhibit ecdysis when applied exogenously to insects.
For example, ponasterone A, which differs from ecdysterone only in
the absence of a hydroxyl group at the C-25 position (Figure 1),
inhibited the ecdysis of <u>Cecropia</u> silkworms (10), <u>Tribolium</u>
<u>confusum</u> (confused flour beetle) larvae (11), and <u>Pectinophora</u>
<u>gossypiella</u> (pink bollworm) larvae (12). Ponasterone A, isolated
from the conifer <u>Podocarpus</u> <u>gracilior</u> (Podocarpaceae), fed in
artificial diet to <u>P</u>. <u>gossypiella</u> larvae resulted in insects each
with three head capsules because they underwent two failed molting
cycles before death (Figure 3). Even though feeding became
impossible after the first inhibited ecdysis, because the adhering
second head capsule covered the mouthparts, the larva synthesized
a third head capsule before death . Ecdysis was inhibited in the
insect to which ponasterone A, a molting hormone analog, was
administered because a rapid decline of the ecdysteroid titre
below a certain critical level is necessary to trigger the release
of eclosion hormone which in turn triggers ecdysis (7).

More potent dietary effects on insects were found with
synthetic steroid analogs, such as Δ^7-5β-cholestene-2β,3β,14α-
triol-6-one, which inhibited ecdysis in <u>Blattella</u> <u>germanica</u> (German
cockroach), and sclerotization and metamorphosis in <u>Musca</u> <u>domestica</u>
(house fly) and <u>T</u>. <u>confusum</u> (13). Other synthetic analogs, the
azasterols, inhibited the Δ^{24} and $\Delta^{22,24}$ sterolreductase
enzymes in <u>Manduca</u> <u>sexta</u> (tobacco hornworm) and the growth and
development of <u>Aedes</u> <u>aegypti</u> (yellow-fever mosquito), <u>T</u>. <u>confusum</u>,
and <u>M</u>. <u>domestica</u> (14). None of the ecdysones or their analogs
have been commercially exploited for insect control because of
their limited efficacy when topically applied or fed to many
species of economically important insects and because of the
complexity of the steroid nucleus (15).

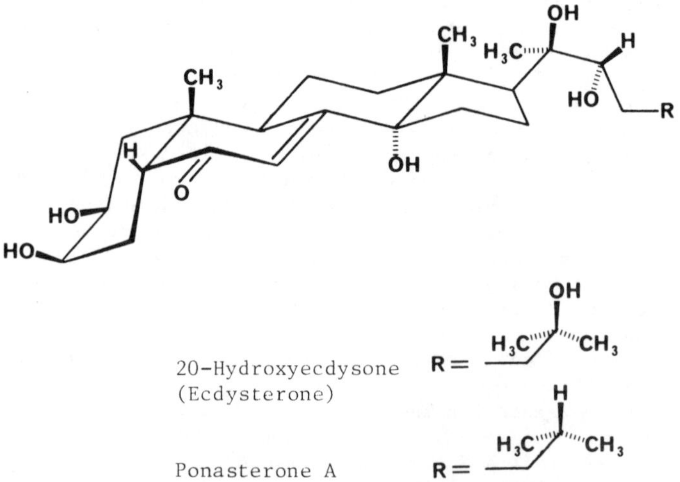

20-Hydroxyecdysone   R =
(Ecdysterone)

Ponasterone A   R =

Figure 1. Stereostructures of the insect molting hormone, ecdysterone and the phytoecdysone, ponasterone A.

Figure 2. Ecdysis inhibition of a larva of the silkworm, Bombyx mori. The insect underwent normal apolysis, but failed to complete ecdysis and could not remove its head capsule or trunk exuviae. Magnification X 12.

Synthetic analogs of other insect hormones (i.e., juvenile hormones) which inhibit insect ecdysis have been commercialized for the control of mosquitoes, flies, fleas and some stored-products pests (16). Juvenile hormones (JH), a group of three closely related terpenoids of which JH 3 predominates in insects (Figure 4), can antagonize the function of ecdysterone in some insects which can result in ecdysis inhibition (17). In a number of lepidopteran crop pests, exposure to analogs of juvenile hormone (e.g., methoprene, Figure 4) can also result in a delayed and usually abortive ecdysis (17). Similar effects have been observed on the endoparasitic wasps Apanteles congregatus (18) and Nasonia vitripennis (19), and on the small fruit fly, Drosophila melanogaster (20). A derivative of farnesoic acid, known to have JH-like activity in silkworms, has been shown to inhibit ecdysis of A. aegypti and Culex pipiens (house mosquito) larvae (21).

One hypothesis on the mechanism of action of JH in the inhibition of ecdysis implicates the active role of JH in the regulation of ecdysone titres in the insect body (17), possibly a consequence of an effect on the prothoracic glands (22). Another hypothesis implicates an effect of JH on eclosion hormone or its action (20, 18).

Other groups of ecdysis inhibitors act by inhibiting chitin synthesis. The most prominent of these are the synthetic 2,6-dihalogenated benzoyl-phenylurea compounds exemplified by the commercial insecticide, Dimilin (diflubenzuron) (Figure 5) (23, 24). Dimilin acts by interfering with the normal deposition of the endocuticle causing it to be loosely attacked or detached from the epidermal layer (25). The interference of Dimilin in cuticle deposition, controlled by the activity rate of chitin synthetase (UDP-N-acetylglucosamine: chitin N-acetylglucosaminyl transferase), results in ecdysis inhibition and the death of affected insects (26, 27). A similar effect was caused by the fungicidal compound polyoxin D, a peptidyl pyrimidine nucleoside antibiotic produced by Streptomyces cacoi var. asoensis (28, 29). Polyoxin D was found to be a competitive inhibitor of chitin synthetase in fungi (30) and in insects (25, 27, 29).

A natural plant compound that was found to inhibit ecdysis through an inhibition of chitin synthetase is the naphthoquinone plumbagin (Figure 6) (31, 32). Plumbagin, isolated from the roots of the tropical medicinal shrub, Plumbago capensis (Plumbaginaceae), inhibited the ecdysis of four lepidopteran species of agricultural pests (Table I). The larvae of P. gossypiella were the most susceptible since 90% died in the pharate condition during the first molt (EI_{90}) from the ingestion of 400 ppm plumbagin in artificial diet. The EI_{90} value of plumbagin against the larvae of Heliothis zea (corn earworm), H. virescens (tobacco budworm), and Trichoplusia ni (cabbage looper) was approximately 1400 ppm. Concentrations of plumbagin lower than those causing ecdysis inhibition caused growth inhibition. The effective concentration for 50% growth inhibition (EC_{50}) was lowest for P. gossypiella (EC_{50} = 150 ppm) and higher for the other species (EC_{50} = 325 - 350 ppm) (Table I).

In order to test for the possibility that the ecdysis inhibitory activity of plumbagin was due to an inhibition of chitin synthetase, an in vitro assay was employed. Chitin

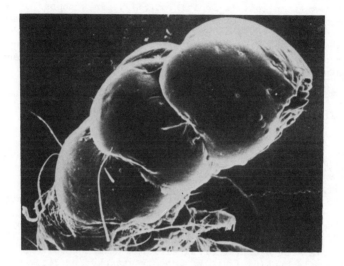

Figure 3. Ecdysis inhibition of a larva of the pink bollworm, <u>Pectinophora gossypiella</u>. The larva three head capsules which masked the functional mouthparts. The insect eventually starved to death. Magnification X 125.

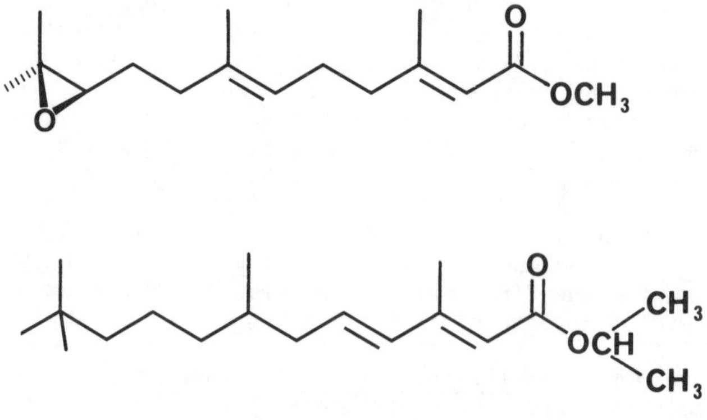

Figure 4. Structures of the insect hormone, juvenile hormone 3 (JH 3) and a commercial analog of JH 3, methoprene.

synthetase extracted from T. ni integuments was preincubated with dithiothreitol for 15 min. The ^{3}H-chitin synthesized from UDP-^{3}H-N-acetylglucosamine (for 60 min at 21° C) was collected on glass microfiber filters and the radioactivity measured. The results in Table II indicate that plumbagin inhibited about 30% of the chitin synthetase extracted from T. ni integuments at a concentration of 3 x 10^{-4} M. This chitin synthetase inhibitory activity is almost comparable to that activity of polyoxin D (Table II). The integumental defect caused by the lowered chitin synthetase activity resulted in a cuticle which could not provide adequate support for the muscular activity and high hemolymph pressure required for normal ecdysis to occur (33).

Attempts were made to enhance the ecdysis inhibitory activity of plumbagin (32). Two analogs of plumbagin based on alkyl elongation at C-2 (n-pentyl and n-decyl) were synthesized and tested. Neither of these analogs showed ecdysis inhibitory activity. Absence of the methyl group at C-2, such as found in the natural plant compound juglone, also resulted in a loss of ecdysis inhibitory activity (31). Thus, while a methyl group is necessary at the C-2 position for activity, elongation of this position with an alkyl side chain resulted in a loss of activity.

Acetylation of the hydroxyl group at C-5 of plumbagin had no effect on the biological activity. However, elongation of the ester group at this position resulted in a loss of ecdysis inhibitory activity. Thus, plumbagin and its acetate were more potent in our bioassays than were the synthetic analogs (32).

Although assays with other insects should be conducted, it does not seem from an economic standpoint that plumbagin and its acetate themselves are of sufficient potency to warrant adaptation into pest management strategies. However, the ease of synthesis of plumbagin (34) and its mode of action make it a possible model for synthetic pesticide research. Although the limited attempts to enhance the activity of plumbagin have been thus far unsuccessful, additional work will hopefully result in enhanced activity.

Already discussed are three types of ecdysis inhibitors, ecdysterone and its analogs, JH and its analogs (juvenoids), which antagonize the action of ecdysterone, and phenylurea compounds (e.g., Dimilin) and polyoxin D, which act by inhibiting chitin synthetase. Other chemicals inhibit ecdysis by different, yet unknown, mechanisms. Thiosemicarbazones inhibited ecdysis when topically applied to fifth instar Oncopeltus fasciatus (large milkweed bug) (35). Similar to the effect found with juvenoids on B. germanica nymphs and Pieris brassicae (cabbage butterfly) larvae (17), thiosemicarbazones delayed the onset of the ecdysteroid peak in O. fasciatus (35), yet dissimilar to the former, restoration of the ecdysteroid titre by injection did not prevent the molting aberrations. Topical application of a juvenoid to fifth instar O. fasciatus resulted in an accelerated production of ecdysteroid, and a shortened molting period (22), opposite to the effects found with the thiosemicarbazones. The thiosemicarbazones had no chitin synthesis-inhibiting effects in an in vitro system which is highly sensitive to Dimilin (35). Possibly, the thiosemicarbazones affect the neurosecretory mechanism by which ecdysterone is regulated.

A natural plant compound which was found to inhibit ecdysis

Table I. Inhibition of ecdysis and growth in four species of
lepidopteran larvae fed plumbagin in an artificial
diet. Assay period was for 10 days in a dark
incubator at 25°C and 80% RH

Species	Instar	EI_{90} (ppm)[a]	EC_{50} (ppm)[b]
Pectinophora gossypiella (Gelechiidae)	first	400 $\pm$ 55	150 $\pm$ 35
Heliothis zea (Noctuidae)	first	1400 $\pm$ 225	350 $\pm$ 50
H. virescens	first	1300 $\pm$ 195	325 $\pm$ 50
Trichoplusia ni (Noctuidae)	second	1400 $\pm$ 250	350 $\pm$ 75

a) EI_{90} values are the concentrations (ppm) of plumbagin causing
90% ecdysis inhibition = death and they are the means of 3
determinations $\pm$ SE with 30 larvae/determination.
b) EC_{50} values are the concentrations (ppm) of plumbagin causing
50% growth inhibition and they are the means of 3 determinations
$\pm$ SE with 40 larvae/determination.

Table II. Inhibition of chitin synthetase extracted from fourth-
instar Trichoplusia ni integuments. 100 μl of enzyme
suspension were preincubated for 15 min at 21°C with
200 μl tris-HCl buffer (pH 7.25), 10mM $MgCl_2$, 1mM
DTT, 18 mM GlcNAc, and either plumbagin dissolved in 10
μl DMSO or simply 10 μl pure DMSO. Reaction initiated
at room temperature with 0.3 μCi UDP-^{3}H-GlcNAc.
Following 60 min incubation (room temperature), reaction
stopped by 5 ml 5% TCA. Acid-insoluble material was
collected on glass-fiber filters, washed repeatedly with
TCA, dried and radioassayed

Inhibitor	Inhibitor Concentration (M)	Chitin Synthetase Inhibition (as % of control)
Plumbagin	3×10^{-6}	1.0
	3×10^{-5}	4.3
	3×10^{-4}	28.7
Polyoxin D	3×10^{-4}	31.3

in insects, by some unknown mechanism, is azadirachtin (Figure 7),
a tetranortriterpenoid of the limonoid type isolated from the
seeds of the neem tree, Azadirachta indica (Meliaceae) (36) and
the fruits of the chinaberry tree, Melia azedarach (Meliaceae)
(37). Although the skeletal structure and stereochemistry of
azadirachtin have not been totally resolved (Figure 7) (38), the
potent ecdysis inhibitory activity of this compound is well known
(39, 40, 41). The ecdysis inhibitory activity of azadirachtin fed
in artificial diet to three species of agricultural pests is shown
in Table III. Azadirachtin was the most active against Spodoptera
frugiperda (fall armyworm), the effective concentration for 95%
ecdysis inhibition (EI_{95}) being 1 ppm in artificial diet. The
ecdysis inhibitory activity of azadirachtin against H. zea and P.
gossypiella was 2 ppm and 10 ppm, respectively. Concentrations of
azadirachtin lower than those causing ecdysis inhibition (i.e.,
less than 1 ppm) inhibited the growth of the treated larvae.
Thus, as a growth inhibitor, azadirachtin was more potent against
S. frugiperda and P. gossypiella, the effective concentration for
50% growth inhibition (EC_{50}) being 0.4 ppm, than against H. zea
($EC_{50} = 0.7$ ppm) (Table III).

Another natural product which was isolated from A. indica and
found to have potent ecdysis inhibitory activity is deacetylazadirachtinol (Figure 8) (42). The biological activity of
deacetylazadirachtinol fed in artificial diet to H. virescens is
shown in Table IV. Although deacetylazadirachtinol was about
2.5-fold less active than was azadirachtin as an insect growth
inhibitor ($EC_{50} = 0.17$ and 0.07 ppm, respectively), the two
compounds had the same ecdysis inhibitory activity ($EI_{50} =
0.80$ ppm).

Comparison of the structure (Figure 7) and activity (Table IV)
of azadirachtin with those of deacetylazadirachtinol (Figure 8 and
Table IV) seems to indicate that the ketal in the C-ring of
azadirachtin is not necessary for potent ecdysis inhibitory
activity, at least against H. virescens. This may be analogous to
the observation of Dreyer (1983) that many of the most potent of
the insect feeding deterrent limonoids are either of the C-ring
secotype or with features which allow C-ring opening.

Several derivatives of azadirachtin were prepared and tested
for biological activity. 3-Deacetylazadirachtin was found to have
the same ecdysis inhibitory activity as that found with azadirachtin against H. virescens, while 13-acetylazadirachtin was found to
be about one-half as active in the same bioassay.

A number of other limonoids which were isolated from plants
in the Meliaceae and the Rutaceae were tested for ecdysis and
growth inhibitory activities (Table V). Cedrelone (Figure 9) was
unique among the compounds tested in Table V since it was the only
limonoid, besides azadirachtin, whose toxicity ($LC_{50} = 150$ ppm)
included ecdysis inhibition when fed to pink bollworm larvae (43).

Although it is unknown how azadirachtin acts as an ecdysis
inhibitor, it apparently does not act by inhibiting chitin
synthetase (32). Possibly, ecdysis inhibition is caused by a
disruption of the titre of ecdysterone (40, 44) or by an
interference with the neuroendocrine system, prothoracicotropic
hormone and allatotropic hormone, which controls the titres of
molting hormone and juvenile hormone, respectively (45).

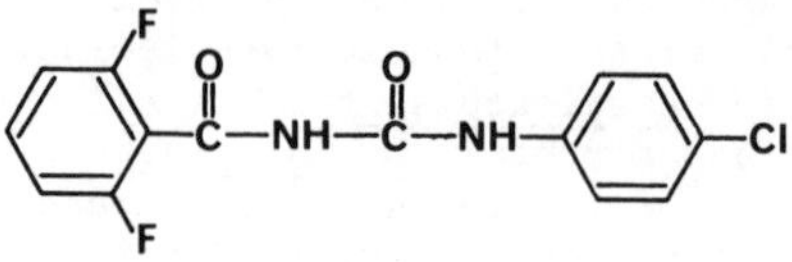

Figure 5.   Dimilin, insect ecdysis and chitin synthetase inhibitor.

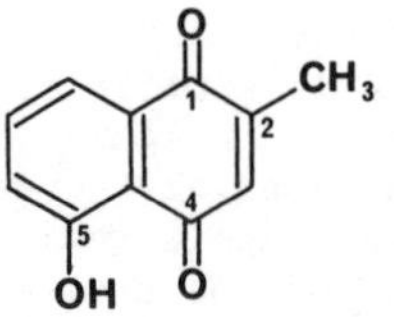

Figure 6.   Plumbagin, insect ecdysis and chitin synthetase inhibitor isolated from _Plumbago_ _capensis_ (Plumbaginaceae).

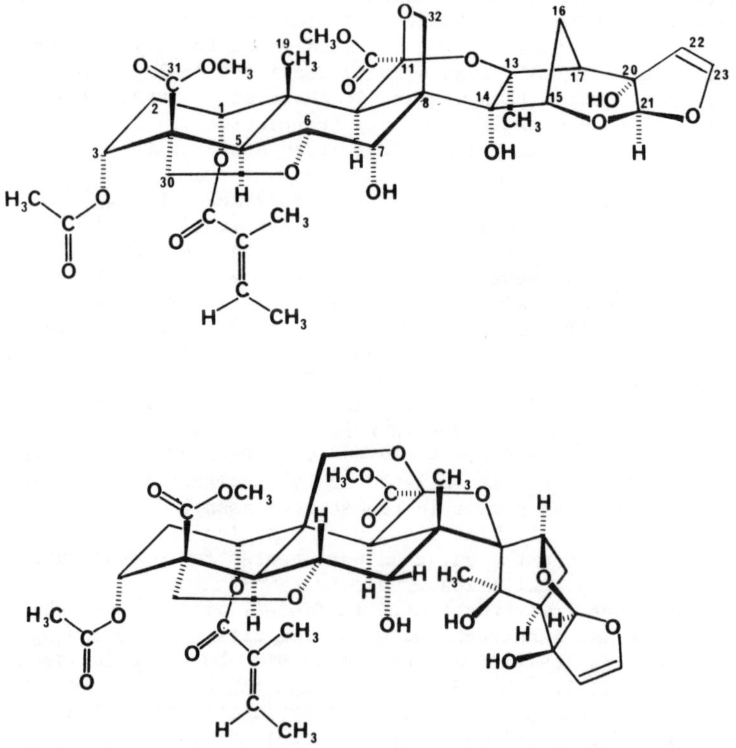

Figure 7.   Upper: previously accepted stereostructure of azadirachtin (36); Lower: recently proposed stereostructure of azadirachtin (38).

Table III.   Inhibition of ecdysis and growth in three species of
first-instar lepidopteran larvae fed azadirachtin in
an artificial diet. Assay period was 10 days in a
dark inculator at 25°C and 80% RH

Species	EI_{95} (ppm)[a]	EC_{50} (ppm)[b]
Heliothis zea	2	0.7
Spodoptera frugiperda	1	0.4
Pectinophora gossypiella	10	0.4

a) EI_{95} values are the concentrations (ppm) of azadirachtin
   causing 95% ecdysis inhibition.
b) EC_{50} values are the effective concentrations (ppm) of
   azadirachtin causing 50% growth inhibition.

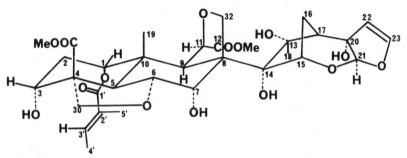

Figure 8.   Stereostructure of the insect ecdysis inhibitory
limonoid deacetylazadirachtinol. Skeleton based on the
structure of azadirachtin proposed by Zanno et al. (36).

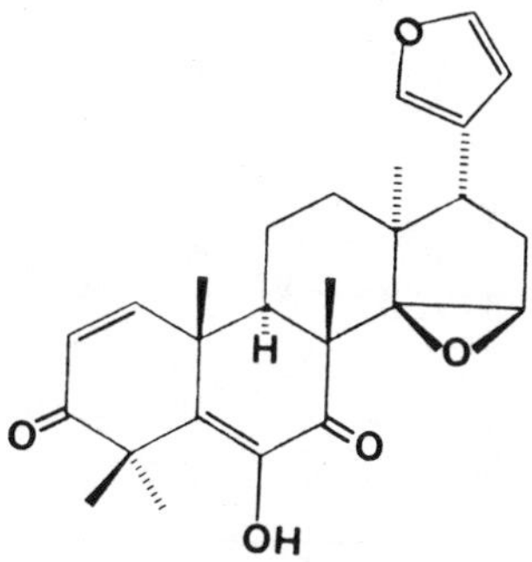

Figure 9.   Structure of the limonoid, cedrelone, the toxicity
of which included insect ecdysis inhibition.

Table IV. Inhibition of ecdysis and growth in first-instar
Heliothis virescens larvae fed two neem oil limonoids
in an artificial diet. Assay period was 10 days in a
dark incubator at 25°C and 80% RH

Test Compound	$EI_{50}{}^a$ (ppm)	$EC_{50}{}^b$ (ppm)	95%[c] Confidence Limits
Deacetylazadirachtinol		0.17	0.12-0.23
	0.80		0.66-0.97
Azadirachtin		0.07	0.05-0.10
	0.80		0.46-1.39

a) EI_{50} values are the concentrations (ppm) causing 50% ecdysis
   inhibition.
b) EC_{50} values are the concentrations (ppm) causing 50% growth
   inhibition.
c) The 95% confidence limits were determined using the methods of
   Litchfield and Wilcoxon (46).

Table V. Insect growth inhibitory activity of some Meliaceae and
Rutaceae limonoids. Values are the dietary concentrations
for 50% growth inhibition (EC_{50})

	Insect Species		
Limonoid	Pectinophora gossypiella	Spodoptera frugiperda	Heliothis zea
Azadirachtin	0.4	0.4	0.7
Cedrelone	3	2	8
Anthotheol	8	3	24
Sendanin	9	11	45
Methylangolensate	15	40	60
Nkolbisonin	20	65	71
7-Deacetylgedunin	22	60	165
Gedunin	32	47	50
Azadiradione	42	130	250
7-Ketogedunin	51	800	900
Nomilin	–	72	95
Obacunone	–	70	97
Evodoulone	96	120	80
7-Deacetylproceranone	175	350	740
Tecleanine	210	320	–
Limonin	–	756	900
7-Deacetylazadiradione	290	5000	3500
Deactylnomilin	950	NE[*]	NE[*]

*NE = no effect at 2000 ppm.

The mode of action of chemicals, besides those inhibiting chitin synthetase, which inhibit ecdysis is difficult to ascertain. Predominantly this is due to the complexity of the molting process and the number of hormones involved in its coordination. The effects of several of these hormones (i.e., ecdysterone and JH) vary depending on the exogenous dose applied, the endogenous titre at the time of application, the species and stage of the treated insect, etc. Nevertheless, the use of ecdysis inhibitors in insect control is possible since the endpoint effect, that of ecdysis inhibition and concomitant death, is easily observed as the most visually dramatic aspect of the molting process (7). Their use in insect control is desirable since ecdysis inhibitors are specific for chitin bearing organisms. Thus, it might be expected that both natural (e.g., phytoecdysones, juvenoid hormones, polyoxin D, azadirachtin) and synthetic (e.g., dimilin, thiosemicarbazones) ecdysis inhibitors will play an increasingly important role in insect control strategies.

Acknowledgments

The authors thank Dr. M. F. Balandrin for helpful discussions. This work was supported in part by a grant awarded by the U.S. National Science Foundation (PCM-8314500).

Literature Cited

1. Cottrell, C. B. Adv. Insect Physiol. 1964, 2, 175–218.
2. Brunet, P. C. In "Aspects of Insect Biochemistry"; Goodwin, T. W., Ed.; Academic Press: New York, 1966; pp. 49–77.
3. Wyatt, G. R. In "Biochemical Actions of Hormones"; Litwack, G., Ed.; Academic Press: New York, 1972; pp. 386–490.
4. Gilbert, L. I.; King, D. In "Physiology of Insecta"; Rockstein, M., Ed.; Academic Press: New York, 1973; pp. 249–370.
5. Meglitsch, P. A. "Invertebrate Zoology"; Oxford University Press: London, 1972; p. 460.
6. Truman, J. W.; Taghert, P. H.; Copenhaver, P. F.; Tublitz, N. J.; Schwartz, L. M. Nature (London) 1981, 291, 70-1.
7. Truman, J. W. Amer. Zool. 1981, 21, 655-61.
8. Hetru, C.; Horn, D. H. S. In "Progress in Ecdysone Research"; Hoffman, J. A., Ed.; Elsevier/North Holland Biomedical Press: New York, 1980; pp. 13–28.
9. Hikino, H.; Takemoto, T. In "Invertebrate Endocrinology and Hormonal Heterophylly"; Burdette, W. J., Ed.; Springer-Verlag: New York, 1974; pp. 185–203.
10. Riddiford, L. M. In "Chemical Ecology"; Sondheimer, E.; Simeone, J.B., Eds.; Academic Press: New York, 1970, p. 114.
11. Robbins, W. E.; Kaplanis, J. N.; Thompson, M. J.; Shortino, T. J.; Joyner, S. C. Steroids 1970, 16, 105-25.
12. Kubo, I.; Klocke, J. A. In "Plant Resistance to Insects"; Hedin, P. A., Ed.; ACS SYMPOSIUM SERIES No. 208, American Chemical Society: Washington, D.C., 1983; pp. 329–46.
13. Robbins, W. E.; Kaplanis, J. N.; Thompson, M. J.; Shortino, T. J.; Cohen, C. F.; Joyner, S. C. Science 1968, 161, 1158-60.

14. Thompson, M. J.; Serban, N. N.; Robbins, W. E.; Svoboda,
     J. A.; Shortino, T. J.; Dutky, S. R.; Cohen, C. F. Lipids
     1974, 10, 615-22.
15. Nakanishi, K. In "Natural Products and the Protection of
     Plants"; Marini-Bettolo, G. B., Ed.; Elsevier Scientific
     Publ. Co.: New York, 1977; pp. 185-210.
16. Staal, G. B. Ann. Rev. Entomol. 1975, 20, 417-60.
17. Vogel, W.; Masner, P.; Graf, O.; Dorn, S. Experientia 1979,
     35, 1254-6.
18. Beckage, N. E.; Riddiford, L.M. J. Insect Physiol. 1982, 28,
     329-34.
19. De Loof, A.; Van Loon, J.; Hadermann, F. Ent. exp. & appl.
     1979, 26, 301-13.
20. Madhavan, K. J. Insect Physiol. 1973, 19, 441-53.
21. Spielman, A.; Skaff, V. J. Insect Physiol. 1967, 13, 1087-95.
22. Smith, W. A.; Nijhout, H. F. J. Insect Physiol. 1981, 27,
     169-73.
23. Van Daalen, J. J.; Meltzer, J.; Mulder, R.; Wellinga, K.
     Naturwissenschaften 1972, 59, 312-3.
24. Wellinga, K.; Mulder, R.; Van Daalen, J. J. J. Agric. Food
     Chem. 1973, 21, 348-54.
25. Salama, H. S.; Motagally, Z. A.; Skatulla, U. Z. angew. Ent.
     1976, 80, 396-407.
26. Mulder, R.; Gijswijt, M. J. Pestic. Sci. 1973, 4, 737-45.
27. Sowa, B. A.; Marks, E. P. Insect Biochem. 1975, 5, 855-9.
28. Endo, A.; Kakiki, K.; Misato, T. J. Bacteriol. 1970, 104,
     189-96.
29. Gijswijt, M. J.; Deul, D. H.; De Jong, B. J. Pestic. Biochem.
     Physiol. 1979, 12, 87-94.
30. Endo, A.; Misato, T. Biochem. Biophys. Res. Comm. 1969, 37,
     718-22.
31. Kubo, I.; Uchida, M.; Klocke, J. A. Agric. Biol. Chem. 1983,
     47, 911-3.
32. Kubo, I.; Klocke, J. A.; Matsumoto, T.; Kamikawa, T. In
     "IUPAC, Pesticide Chemistry: Human Welfare and the Environ-
     ment, Rational and Biorational Design of Pesticides and
     Growth Regulators"; Miyamoto, J., Ed.; Pergamon Press: New
     York, 1983; pp. 169-75.
33. Chapman, R. F. "The Insects, Structure and Function";
     American Elsevier Publ. Co., Inc.: New York, 1979; p. 700.
34. Ichihara, A.; Ubrukata, M.; Sakamura, S. Agric. Biol. Chem.
     1980, 44, 211-3.
35. Kelly, T. J.; Redfern, R. E.; DeMilo, A. B.; Borkovec, A.B.
     Pestic. Biochem. Physiol. 1982, 17, 35-41.
36. Zanno, P. R.; Miura, I.; Nakanishi, K.; Elder, D. L. J. Am.
     Chem. Soc. 1975, 97, 1975-7.
37. Butterworth, J. H.; Morgan, E. D. J. Chem. Soc. 1968, Chem.
     Commun. 23-4.
38. Taylor, R. B.; Tempesta, M. S. Abstract No. 177,
     International Research Congress on Natural Products, Chapel
     Hill, North Carolina, July, 1985.
39. Qadri, S. S. H.; Narsaiah, J. Indian J. Exp. Biol. 1978, 16,
     1141-3.
40. Sieber, K.-P.; Rembold, H. J. Insect Physiol. 1983, 29,
     523-7.

41. Kubo, I.; Klocke, J. A. <u>Agric. Biol. Chem.</u> 1982, 46, 1951-3.
42. Kubo, I.; Matsumoto, A.; Matsumoto, T.; Klocke, J. A.
     <u>Tetrahedron</u> 1985, in press.
43. <u>Klocke</u>, J. A. Ph.D. Thesis, University of California,
     Berkeley, 1982.
44. Rembold, H.; Klaus-Peter, S. <u>Z. Naturforsch. Sect. C Biosci.</u>
     1981, 36, 466-9.
45. Rembold, H. In "Advances in Invertebrate Reproduction 3";
     Engels, W.; Clark, W. H., Jr.; Fischer, A.; Olive, P. J. W.;
     Went, D. F., Eds.; Elsevier Science Publishers: New York,
     1984; pp. 481-91.
46. Litchfield, J. T. Jr.; Wilcoxon, F. <u>J. Pharmacol. Exp. Ther.</u>
     1949, 96, 99-113.

RECEIVED September 13, 1985

18

The Neem Tree: Natural Resistance Par Excellence

Martin Jacobson

Insect Chemical Ecology Laboratory, USDA, Agricultural Reseach Service, Beltsville, MD 20705

Neem (Azadirachta indica A. Juss.), a subtropical tree, is highly resistant to attack by numerous species of insects and nematodes. The tetranortriter-penoids meliantriol, salannin, and azadirachtin, occurring mainly in the seeds, act as antifeedants, disruptants of insect growth and development, or toxicants. Azadirachtin is effective at dosages as low as 0.1 ppm. Although azadirachtin's complex structure probably precludes its ready synthesis, liquid or dust formulations of ethanol extracts of the seeds have been shown to be nontoxic to warm-blooded animals and nonmutagenic, making them suitable for practical use as pesticides. Registration of a neem formulation for use in the United States on nonfood crops by the Environmental Protection Agency is pending.

Neem, Azadirachta indica A. Juss. (family Meliaceae), also known commonly as "nim" or "margosa," is a subtropical tree native to the arid areas of India, Pakistan, Sri Lanka, Malaya, Indonesia, Thailand, Burma, and East Africa. It is under cultivation in plantations of West Africa and the Caribbean islands, as well as parts of Central and South America. Two neem trees brought from India approximately 40 years ago are thriving in the Miami, Florida area of the United States and a number of the young trees cultivated from seed are doing well in Puerto Rico (1). When fully grown the tree can reach a height of 60 feet with a trunk diameter of 6 feet. It is estimated that approximately 14 million neem trees are under cultivation or grow wild in India alone (2).

The durable wood is used for building furniture and sometimes as firewood in the developing countries, the leaves and twigs are spread among household goods and clothing to repel insects, and the bark, which contains 12-14% tannins, is used for the production of industrial chemicals (3, 4). The pulp of the nonedible yellow fruits is a promising substrate for generating methane gas. Potentially the most valuable part of the neem tree is the seed, which contains up to 40% oil that is used in the developing countries

as fuel for lamps, as a lubricant for machinery, and in the preparation of soaps, toothpaste, cosmetics, pharmaceuticals, disinfectants,
and insecticides. Neem cake (the residue remaining after extraction
of the oil from the seeds) is an excellent fertilizer several times
richer in plant nutrients than manure (5). The tree has long been
known to be free of attack by practically all species of insects,
nematodes, and plant diseases, and it thrives in areas under
climatic conditions that are unsuitable for all but the most hardy
species. A considerable amount of scientific literature exists on
the use of neem as an insect control agent. Of the various tree
parts, the seeds are by far the most effective (for comprehensive
reviews of the pesticidal properties of neem, see references 6
and 7).

Chemistry

A host of tetranortriterpenoids have been isolated from various
parts of the neem tree and, although all have not been tested for
pesticidal properties, three compounds obtained from the seed are
quite active as insect feeding deterrents, toxicants, and(or)
disruptants of growth and development against a large variety of
insect species and nematodes (6). These specific compounds are
meliantriol (I), first isolated in 1967 by Lavie et al. (8),
salannin (I,I) (9), and azadirachtin (III) (10, 11). The highly
complicated structure of azadirachtin was identified by Zanno (12)
and Nakanishi (13). Subsequently, a large series of tetranortriterpenoids and pentanortriterpenoids were isolated from the seed
oil by Kraus et al. (14-18), Lucke et al. (19), Garg and Bhakuni
(20), and Kubo et al. (21). Other components isolated from neem
are beta-sitosterol (22), fatty acids (23), and flavanoids (24).
Neem seed oil possesses an unpleasant, garlicky-type odor,
undoubtedly due to a number of sulfur compounds present.

Compounds I and II have been obtained in discrete crystalline
form but azadirachtin has thus far been obtained only as a white
amorphous powder melting at 154-158°C., despite several attempts to
prepare crystalline derivatives (mainly esters). This is probably
due to the presence of different stereoisomers, as Rembold et al.
(25) recently reported the separation of azadirachtin into at least
4 amorphous stereoisomers (azadirachtins A, B, C, and D) with
similar biological activities. Of these three compounds, azadirachtin is by far most active as a feeding deterrent for insects; it is
effective against several species at a concentration of 0.1 ppm or
less (1). However, Rembold et al. (25) have very recently reported
the separation of azadirachtin into at least 4 amorphous stereoisomers (azadirachtins A, B, C, and D) with similar biological
activities. Morgan (26) has shown that the yield of azadirachtin
extracted from 100 grams of neem seed varies markedly according to
the origin of the seed, ranging from 0.2 g using commercial Indian
seed (27, 28) to 3.5 g using seed from Ghana. However, Ermel et al.,
using thin-layer chromatography on ethanol extracts of seed, found
that the best yields of azadirachtin were obtained from seeds
originating in Togo (6.2 g/100 g) and India (3.5 g/100 g) (29). The
procedure used by our Laboratory at the U.S. Department of Agriculture to determine the azadirachtin content in Indian neem extract
and formulations showed (Table I) that 95% ethanol is the preferred
solvent for crude extraction of the active components (28). However,
Feuerhake (30) reported that methanol with or without methyl-tert-
butyl ether was the most efficient extraction solvent in his hands.
Exposure of azadirachtin to sunlight causes a rapid decrease in
antifeeding potency, which is prevented to some extent (25%) by
mixing with neem, angelica, castor, or calamus oil (31).

Table I. Azadirachtin determined by high-performance liquid
chromatography (HPLC) in neem kernel extracts prepared with various
solvents.

Solvent used	Azadirachtin found (μg/10μL)
Ethanol (95%)	2.80
Methanol-H_2O (85:15)	2.60
Methanol	2.19
Methylene chloride	1.73
Ethyl ether	1.28
Acetone	0.74

Pesticidal Activity

Neem extracts are highly effective not only in preventing or
reducing feeding by a variety of insects but also as a repellent,
inhibitor of growth and development, and sterilant. A comprehensive
review published by Warthen (6) in 1979 lists 20 species of the
order Coleoptera (beetles), 4 species of the order Diptera (flies),
5 species of the order Heteroptera (true bugs), 10 species of the
order Homoptera (aphids and scale insects), 2 species of the order
Isoptera (termites), 30 species of the order Lepidoptera (moths and
butterflies), and 7 species of the order Orthoptera (locusts, grass-

hoppers, crickets) affected by neem preparations as feeding inhibitors or growth regulators through the year 1978. Also included were
5 species of nematodes and a species of mite (Acari) whose development is disrupted by neem.

Table II lists those species evaluated for feeding inhibition
(FI), growth and development regulation (GR), and toxicity (T) with
neem since 1978. Included are 2 species of mites, 25 species of
Coleoptera, 10 species of Diptera, 3 species of Heteroptera, 6
species of Homoptera, 1 species of Hymenoptera, 25 species of
Lepidoptera, and 9 species of Orthoptera. The total number of insect
species known to be adversely affected by neem preparations is
presently 123, in addition to 3 species of mites and 5 species of
nematodes. There is every reason to believe that these numbers will
increase rapidly, judging by the large number of scientists now
involved in neem research.

Pharmacological Activity

Malaria-infected mice fed daily oral doses of an ethanol extract of
neem leaves failed to recover and died from the disease after the
fifth day (119). The extract caused an appreciable contraction of
guinea pig ileum, and intravenous injection in the dog resulted in
an initial rise in blood pressure followed shortly by a protracted
fall; the effect was similar to that of histamine (120). A methanol
extract of the combined leaves and bark of neem had a pronounced
anti-inflammatory effect on rat paw and an antipyretic effect in
rabbits (121). Injection of an aqueous extract of the leaves into
guinea pigs and rabbits profoundly reduced blood pressure and
reduced the heart rate (122).

Anti-ulcer studies conducted with laboratory rats using
nimbidin, the major bitter principle of neem seed oil, showed
significant anti-ulcer potential when fed at doses of 20-40 mg/kg.
The compound also afforded remarkable protection from duodenal
lesions (123).

Neem oil possesses strong in vitro spermicidal action (within
30 sec) against rhesus monkey and human spermatozoa. When used
intravaginally at 20 L in rats and at 1 mL in monkeys and in human
subjects the oil was completely effective in preventing pregnancy
without causing side effects (124). The sodium salt of nimbidin
also acts as a spermicide (125). Single or multiple intravaginal
application of neem oil during the post-implantation period prevented pregnancy in rats, but normal fertility was restored after
30 days of withdrawal (126).

Toxicological Activity

Birds eat the neem fruit in large quantity with impunity but spit
out the seed, which has a strongly bitter taste (2). An ethanol
extract of the seed instilled into the rabbit eye caused no irrita-
tion and no significant skin-sensitizing reaction resulted from
injection into the shaved skin of a guinea pig. The standard Ames
test with azadirachtin showed no mutagenic activity on four strains
of Salmonella typhimurium (127). The acute oral toxicity of the
extract in mice was extremely low (13 g/kg) (121).
Incorporation of neem seed cake (ethanol-extracted and unextracted)
into the ration of lambs at 10% caused no change in the growth rate,

Table II. Arthropods and nematodes evaluated with neem for feeding inhibition (FI), growth regulation (GR), and toxicity (T) (1979-1984).

Scientific name	Common name	Stage	Activity	Reference
ACARI				
Tetranychus cinnabarinus (Boisd.)	carmine spider mite	adult	FI, GR	32
Tetranychus urticae Koch.	2-spotted spider mite	adult	FI, GR	33
COLEOPTERA				
Acalymma vittatum (Fabricius)	striped cucumber beetle	adult	FI	34, 35
Callosobruchus chinensis Linnaeus	pulse beetle	adult	FI	36, 37
Callosobruchus maculatus (Fabricius)	cowpea weevil	larva, adult	T, FI	38-40
Chirida bipunctata Fabricius	tortoise beetle	adult	FI	36
Diabrotica undecimpunctata howardi Barber	spotted cucumber beetle	adult	FI	1, 41
Diacrysia obliqua Walker	jute hairy caterpillar	larva	FI	37
Dicladispa armigera Oliv.	rice hispa	adult	FI	37
Epilachna chrysomelina (Fabricius)	- - - -	larva	FI	42
Epilachna punctata Mulsant	12-spotted ladybird beetle	adult	FI	42
Epilachna varivestis Mulsant	Mexican bean beetle	larva	FI, GR	25, 43-49
		adult	T	50, 51
Euproctis fraterna Moore	- - - -	larva	FI, GR	36
Henosepilachna elaterii Rossi	- - - -	larva	GR	51
Leptinotarsa decemlineata (Say)	spotted cucumber beetle	larva, adult	FI, GR	41, 48
Nephantis serinopa Meyr.	Colorado potato beetle	larva	FI	36
Phyllotreta striolata (Fabricius)	striped flea beetle	adult	FI	52
Podagrica sjostedti Jacoby	flea beetle	adult	FI	42
Podagrica uniforma Jacoby	flea beetle	adult	FI	42
Popillia japonica Newman	Japanese beetle	larva, nymph	GR	53
		adult	FI	54

Scientific name	Common name	Stage	Activity	Reference
Rhyzopertha dominica (Fabricius)	lesser grain borer	adult	FI	41, 55-58
Sitophilus granarius (Linnaeus)	granary weevil	adult	FI	56, 57
Sitophilus oryzae (Linnaeus)	rice weevil	adult	FI, GR, T	58, 59
Sitophilus zeamais Motschulsky	maize weevil	adult	FI	60
Tribolium castaneum (Herbst)	red flour beetle	adult	FI, T	41, 56-57, 61
Tribolium confusum Jacqueline du Val	confused flour beetle	adult	FI, GR, T	49, 60
Trogoderma granarium Everts	khapra beetle	larva, adult		
DIPTERA				
Aedes aegypti (Linnaeus)	yellowfever mosquito	larva	GR	1, 63, 64
Atherigona soccata Rondani	sorghum shoot fly	adult	FI	36
Ceratitis capitata (Wiedemann)	Mediterranean fruit fly	larva, adult	GR, T	65
Culex pipiens Linnaeus	northern house mosquito	larva	T	66
Dacus ciliatus Lw.	fruit fly	larva	(-) FI	51
Dacus dorsalis Hendel	oriental fruit fly	adult	T	67
Liriomyza sativae Blanchard	vegetable leafminer	larva, egg, adult	FI, T, GR	68-70
Liriomyza trifolii (Burgess)	leafminer	larva, adult	FI, GR	68-71
		larva, pupa	FI, T	72
Musca autumnalis de Geer	face fly	larva, adult	T, GR	73, 74
Orseolia oryzae (Wood-Mason)	rice gall midge	adult	T	75
HETEROPTERA				
Dysdercus cingulatus Fabricius	- - - - -	nymph	GR, T	76
Dysdercus koenigii Fabricius	red cotton bug	larva, egg	GR, T	77
Oncopeltus fasciatus (Dallas)	large milkweed bug	adult	T	78
		nymph	GR	79

Continued

Table II—Continued

Scientific name	Common name	Stage	Activity	Reference
HOMOPTERA				
Acyrthosiphon pisum Harr.	aphid	nymph	GR, T	80
Aphis fabae Scopoli	bean aphid	nymph	T	80
Bemisia tabaci (Gennadius)	sweetpotato whitefly	nymph, adult	FI	51
Nephotettix virescens (Distant)	green leafhopper	adult	FI	81–84
Nilaparvata lugens (Stal.)	brown planthopper	larva, nymph, adult	GR, FI	37, 75, 82, 85–87
Sogatella furcifera (Horvath)	whitebacked planthopper	adult	FI, T	82, 88
HYMENOPTERA				
Apis mellifera Linnaeus	honey bee	larva	GR, (−)FI	89
		adult	FI	41
LEPIDOPTERA				
Anagasta kuhniella (Zeller)	Mediterranean flour moth	larva	FI, GR, T	89–92
Cadra cautella (Walker)	almond moth	larva, pupa	GR, T	58
Cnaphalocrocis medinalis (Guenee)	rice leaf folder	larva	FI	93, 94
		adult	GR	94
Crocidolomia binotalis Zeller	cabbage webworm	larva	FI, GR, T	95–97
Cydia pomonella (Linnaeus)	codling moth	larva	GR, T	41
Diacrisia obliqua Walker	jute hairy caterpillar	larva	FI	37
Diaphania hyalinata (Linnaeus)	melonworm	larva	FI, T	41
Earias insulana (Boisduval)	spiny bollworm	larva	FI	98
Heliothis armigera (Hubner)	cotton bollworm	larva, adult	FI, T	99, 100
Heliothis virescens (Fabricius)	tobacco budworm	larva	FI	41, 99, 101
Heliothis zea (Boddie)	corn earworm, bollworm	larva	FI	101
Lymantria dispar (Linnaeus)	gypsy moth	larva	FI	41
Mamestra brassicae Linnaeus	cabbage armyworm	larva	FI	99

Scientific name	Common name	Stage	Activity	Reference
Manduca sexta (Linnaeus)	tobacco hornworm	larva	FI, GR, T	102
Mythimna separata (Walker)	rice earcutting caterpillar	larva	GR	93, 103
Papilio demodocus Esper	– – – –	larva	FI	42
Pectinophora gossypiella (Saunders)	pink bollworm	larva	FI	101
Phthorimaea operculella (Zeller)	potato tuberworm	larva	FI	104
Plutella xylostella (Linnaeus)	diamondback moth	larva	FI, T	6, 48, 105
Sitotroga cerealella (Olivier)	Anjoumois grain moth	adult	FI, T	55
Spodoptera exempta (Walker)	nutgrass armyworm	larva	FI	99
Spodoptera frugiperda (J. E. Smith)	fall armyworm	larva	FI, GR, T	41, 80, 99, 106, 107
Spodoptera littoralis (Boisduval)	Egyptian cotton leafworm	larva	FI, GR	98-109
Spodoptera litura (Fabricius)	tobacco caterpillar	larva	FI, GR	36, 110, 111
		adult	repellent	112
Trichoplusia ni (Hubner)	cabbage looper	larva	FI	99
ORTHOPTERA				
Acheta domesticus (Linnaeus)	house cricket	nymph, adult	GR	113
Amsacta moorei Butler	– – – –	larva	FI	114
Chortoicetes terminifera (Walker)	plague locust	adult	(-) FI	41
Diapheromera femorata (Say)	walkingstick	adult	FI	115
Dissosteira carolina (Linnaeus)	Carolina grasshopper	adult	FI	115
Gryllus pennsylvanicus Burmeister	field cricket	adult	FI	115
Locusta migratoria migratorioides R. & F.	migratory locust	larva	GR	116
		adult	GR, T	117
Phaulacridium spp.	wingless grasshopper	adult	(-) FI	41
Schistocerca gregaria Forsk	desert locust	adult	FI	118
NEMATODA				
Meloidogyne incognita (Kofoid & White)	root-knot nematode	adult	(-) FI	41

(-) = negative response

20% increased the growth rate, and 30% significantly reduced the
growth rate. Blood glucose, hemoglobin content, and urea nitrogen
remained normal at all dosages (128).

Neemrich 100, a formulation containing 30% neem oil, applied
dermally to albino rats at daily dose levels of 200, 400, and 600
mg/kg for 3 weeks, caused no overt signs of toxicity or abnormal
behavior. Treated rats exhibited higher food consumption, gained
weight, and showed no abnormal blood levels (129).

Albino rats and mice showed no toxicity when fed nimbidin up to
2000 mg/kg or when administered intraperitoneally at 1000 mg/kg.
Dogs fed nimbidin daily at 10 and 20 mg/kg for 4 weeks showed no
systemic toxicity (130).

Malaysian infants given large (up to 5 mL) oral doses of neem
oil for minor ailments showed vomiting, drowsiness, metabolic
acidosis, polymorpho-nuclear leucocytosis, and encephalopathy within
hours of ingestion (131).

Potential for Practical Use as a Pesticide
Homemade formulations of neem seeds have been used effectively
by farmers in the developing countries for many years for controlling agricultural pests of tobacco and cotton (3,7). Although
the compexity of the azadirachtin molecule probably precludes its
ready synthesis, both crude and partially purified extracts of the
seeds can be used for pest control. Improved methods for isolating
highly active fractions have been developed, a neem pilot plant is
presently in operation in Burma (132), and the cultivation and
processing of neem on a large scale are being sponsored in various
parts of the world by the German Agency for Technical Cooperation,
the U.S. Agency for International Development, and several agricultural agencies (1, 4, 5, 133, 134).

The first hurdle toward the commercial development and
marketing of neem formulations for controlling agricultural pests
in the United States is being overcome by a private concern which
is to be granted registration by the Environmental Protection Agency
(EPA) for use on nonfood crops (135). Since a large volume of
toxicological test data has been supplied to EPA showing that the
formulations are nontoxic to man and animals, registration may be
granted for use on food crops as well.

Literature Cited

1. Jacobson, M. Proc. 1st Intern. Neem Conf., 1981, pp. 33-42.
2. Saxena, R. C., personal communication.
3. Radwanski, S. World Crops Livestock 1977, 29, 111-113.
4. "Firewood Crops. Shrub and Tree Species for Energy Production,"
    National Academy of Sciences, 1980, p. 114.
5. Michel-Kim, H.; Brandt, A. Proc. 1st Intern. Neem Conf. 1981,
    pp. 279-290.
6. Warthen, J. D., Jr. USDA Sci. Educ. Admin., Agric. Revs.
    Manuals ARM-NE-4, 1979, 22 pp.
7. Jain, H. K. Indian Agric. Res. Inst. Bull. No. 40, 1983, 63 pp.
8. Lavie, D.; Jain, M. K.; Shpan-Gabrielith, S. R. Chem. Commun.
    1967, 910.
9. Henderson, R.; McCrindle, R.; Overton, K. H. Tetrahedron Lett.
    1964, 3969-3974.
10. Butterworth, J. H.; Morgan, E. D. Chem. Commun. 1968, 23-24.

11. Butterworth, J. H.; Morgan, E. D. J. Insect. Physiol. 1971, 17, 969-977.
12. Zanno, P. R. Ph.D. Thesis, Columbia University, New York, 1974.
13. Nakanishi, K. Recent Adv. Phytochem. 1975, 9, 283-298.
14. Kraus, W.; Cramer, R. Tetrahedron Lett. 1978, 2395-2398.
15. Kraus, W.; Cramer, R.; Sawitzki, G. Phytochemistry 1981, 20, 117-120.
16. Kraus, W.; Cramer, R. Liebigs Ann. Chem. 1981, 181-189.
17. Kraus, W.; Cramer, R. Chem. Ber. 1981, 114, 2375-2381.
18. Kraus, W.; Cramer, R.; Bokel, M.; Sawitzki, G. Proc. 1st Intern. Neem Conf., 1981, pp. 53-62.
19. Lucke, J.; Fuchs, S.; Voelter, W. Planta Med. 1980, 39, 280.
20. Garg, H. S.; Bhakuni, D. S. Phytochemistry 1984, 10, 2383-2385.
21. Kubo, I.; Matsumoto, T.; Matsumoto, A.; Shoolery, J. N. Tetrahedron Lett. 1984, 25, 4729-4732.
22. Quasim, C.; Dutta, N. L. Indian J. Appl. Chem. 1970, 33, 384-386.
23. Skellon, J. H.; Thorburn, S.; Spence, J.; Chatterjee, S. N. J. Sci. Food Agric. 1962, 13, 639-643.
24. Subramanian, S. S.; Nair, A. G. R. Indian J. Chem. 1972, 10, 452.
25. Rembold, H.; Forster, H.; Czoppelt, C.; Rao, P. J.; Sieber, K. P. Proc. 2nd Intern. Neem Conf. 1984, pp. 153-161.
26. Morgan, E. D. Proc. 1st Intern. Neem Conf. 1981, pp. 43-52.
27. Uebel, E. C.; Warthen, J. D., Jr.; Jacobson, M. J. Liquid Chromatog. 1979, 2, 875-882.
28. Warthen, J. D., Jr.; Stokes, J. B.; Jacobson, M.; Kozempel, M. F. J. Liquid Chromatog. 1984, 7, 591-598.
29. Ermel, K.; Pahlich, E.; Schmutterer, H. Proc. 2nd Intern. Neem Conf. 1984, pp. 91-94.
30. Feuerhake, K. J. Proc. 2nd Intern. Neem Conf. 1984, pp. 103-114.
31. Stokes, J. B.; Redfern, R. E. J. Environ. Sci. Health 1982, A17, 57-65.
32. Mansour, F. A.; Ascher, K. R. S. Phytoparasitica 1983, 11, 177-185.
33. Schauer, M.; Schmutterer, H. Proc. 1st Intern. Neem Conf. 1981, pp. 259-265.
34. Reed, D. K.; Jacobson, M.; Warthen, J. D., Jr.; Uebel, E. C.; Tromley, N. J.; Jurd, L.; Freedman, B. USDA-SEA Tech. Bull. 1641, 1981, 13 pp.
35. Reed, D. K.; Warthen, J. D., Jr.; Uebel, E. C.; Reed, G. L. J. Econ. Entomol. 1982, 75, 1109-1113.
36. Abdul Kareem, A. Proc. 1st Intern. Neem Conf. 1981, pp. 223-250.
37. Islam, B. N. Proc. 2nd Intern. Neem Conf. 1984, pp. 263-290.
38. Pereira, J. J. Stored Prod. Res. 1983, 19, 57-62.
39. Zehrer, W. Proc. 2nd Intern. Neem Conf. 1984, pp. 453-460.
40. Ivbijaro, M. F. Prot. Ecol. 1983, 5, 177-182.
41. Jacobson, M.; Stokes, J. B.; Warthen, J. D., Jr.; Redfern, R. E.; Reed, D. K.; Webb, R. E.; Telek, L. Proc. 2nd Intern. Neem Conf. 1984, pp. 31-42.
42. Redknap, R. S. Proc. 1st Intern. Neem Conf. 1981, pp. 205-214.
43. Schmutterer, H.; Rembold, H. Z. Angew. Entomol. 1980, 89, 179-188.
44. Ascher, K. R. S.; Gsell, R. Z. Pflanzenkr. Pflanzenschutz 1981, 88, 764-767.

45. Lange, W.; Schmutterer, H. Z. Pflanzenkr. Pflanzenschutz 1982, 89, 258-265.
46. Feuerhake, K.; Schmutterer, H. Z. Pflanzenkr. Pflanzenschutz 1982, 89, 737-747.
47. Feuerhake, K. J. Proc. 2nd Intern. Neem Conf. 1984, pp. 103-114.
48. Lange, W. Proc. 2nd Intern. Neem Conf. 1984, pp. 129-140.
49. Schulz, W. D. Proc. 1st Intern. Neem Conf. 1981, pp. 81-96.
50. Schulz, W. D.; Schluter, U. Proc. 2nd Intern. Neem Conf. 1984, pp. 237-252.
51. Dreyer, M. Proc. 2nd Intern. Neem Conf. 1984, pp. 435-444.
52. Meisner, J.; Mitchell, B. K. Z. Pflanzenkr. Pflanzenschutz 1982, 89, 463-467.
53. Ladd, T. L., Jr.; Warthen, J. D., Jr.; Klein, M. G. J. Econ. Entomol. 1984, 77, 903-905.
54. Ladd, T. L., Jr. Proc. 1st Intern. Neem Conf. 1981, pp. 149-156.
55. Devi, D. A.; Mohandas, N. Entomon 1982, 7, 261-264.
56. Jilani, G.; Su, H. C. F. J. Econ. Entomol. 1983, 76, 154-157.
57. Malik, M. M.; Naqvi, S. H. M. J. Stored Prod. Res. 1984, 20, 41-44.
58. Pereira, J.; Wohlgemuth, R. Z. Angew. Entomol. 1983, 94, 208-214.
59. Ivbijaro, M. F. Prot. Ecol. 1983, 5, 353-357.
60. Akou-Edi, D. Proc. 2nd Intern. Neem Conf. 1984, pp. 445-452.
61. Sharma, H. L., Vimal, O. P.; Atri, B. S. Indian J. Agric. Sci. 1981, 51, 896-900.
62. Siddig, S. A. Proc. 1st Intern. Neem Conf. 1981, pp. 251-258.
63. Schmutterer, H.; Zebitz, C. P. W. Proc. 2nd Intern. Neem Conf. 1984, pp. 83-90.
64. Zebitz, C. P. W. Entomol. Exp. Appl. 1984, 35, 11-16.
65. Steffens, R. J.; Schmutterer, H. Z. Angew. Entomol. 1982, 94, 98-103.
66. Chavan, S. R. Proc. 2nd Intern. Neem Conf. 1984, pp. 59-66.
67. Sombatsiri, K.; Tigvattanont, S. Proc. 2nd Intern. Neem Conf. 1984, pp. 95-100.
68. Webb, R. E.; Hinebaugh, M. A.; Lindquist, R. K.; Jacobson, M. J. Econ. Entomol. 1983, 76, 357-362.
69. Larew, H. G.; Webb, R. E.; Warthen, J. D., Jr. Proc. 4th Ann. Ind. Conf. Leafminer, 1984, pp. 108-117.
70. Webb, R. E.; Larew, H. G.; Wieber, A. M.; Ford, P. W.; Warthen, J. D., Jr. Proc. 4th Ann. Ind. Conf. Leafminer 1984, pp. 118-127.
71. Lindquist, R. K., personal communication.
72. Fagoonee, I.; Toory, V. Insect Sci. Appl. 1984, 5, 23-30.
73. Gaaboub, I. A.; Hayes, D. K. Environ. Entomol. 1984, 13, 803-812.
74. Gaaboub, I. A.; Hayes, D. K. Environ. Entomol. 1984, 13, 1639-1643.
75. Chiu, S.-F.; Huang, Z.-X; Huang, D.-P; Huang, B.-Q; Xu, M.-C; Hu, M.-Y. South China Agric. Coll. Res. Bull. No. 3, 1984, 32 pp.
76. Abraham, C. C.; Ambika, B. Curr. Sci. (India) 1979, 48, 554-555.
77. Koul, O. Entomol. Exp. Appl. 1984, 36, 85-88.

78. Koul, O. Z. Angew. Entomol. 1984, 98, 221-223.
79. Redfern, R. E.; Warthen, J. D., Jr.; Uebel, E. C.; Mills, G. D., Jr. Proc. 1st Intern. Neem Conf. 1981, pp. 129-136.
80. Schauer, M. Proc. 2nd Intern. Neem Conf. 1984, pp. 141-150.
81. Mariappan, V.; Saxena, R. C. J. Econ. Entomol. 1983, 76, 573-576.
82. Von der Heyde, J.; Saxena, R. C.; Schmutterer, H. Proc. 2nd Intern. Neem Conf. 1984, pp. 377-390.
83. Mariappan, V.; Saxena, R. C. Proc. 2nd Intern. Neem Conf. 1984, pp. 413-429.
84. Mariappan, V.; Saxena, R. C. J. Econ. Entomol. 1984, 77, 519-521.
85. Saxena, R. C.; Liquido, N. J.; Justo, H. D. Proc. 1st Intern. Neem Conf. 1981, pp. 171-188.
86. Saxena, R. C.; Justo, H. D., Jr.; Epino, P. B. J. Econ. Entomol. 1984, 77, 502-507.
87. Chao, S.-C.; Huang, P.-C.; Hu, M.-Y. Acta Entomol. Sin. 1983, 26, 1-8.
88. Saxena. R. C.; Epino, P. B.; Tu, C.-W.; Puma, B. C. Proc. 2nd Intern. Neem Conf. 1984, pp. 403-412.
89. Rembold, H.; Sharma, G. K.; Czoppelt, C. Proc. 1st Intern. Neem Conf. 1981, pp. 121-128.
90. Sharma, G. K.; Czoppelt, C.; Rembold, H. Z. Angew. Entomol. 1980, 90, 439-444.
91. Maurer, G. Proc. 2nd Intern. Neem Conf. 1984, pp. 365-376.
92. Rembold, H.; Sharma, G. K.; Czoppelt, C.; Schmutterer, H. Z. Angew. Entomol. 1982, 93, 12-17.
93. Schmutterer, H.; Saxena, R. C.; Von der Heyde, J. Z. Angew. Entomol. 1983, 95, 230-237.
94. Saxena, R. C.; Waldbauer, G. P.; Liquido, N. J.; Puma, B. C. Proc. 1st Intern. Neem Conf. 1981, pp. 189-204.
95. Fagoonee, I. Proc. 1st Intern. Neem Conf. 1981, pp. 109-120.
96. Fagoonee, I.; Lauge, G. Phytoparasitica 1981, 9, 111-118.
97. Fagoonee, I. Proc. 2nd Intern. Neem Conf. 1984, pp. 211-224.
98. Meisner, J.; Ascher, K. R. S.; Aly, R.; Warthen, J. D., Jr. Phytoparasitica 1981, 9. 27-32.
99. Simmonds, M. S. J.; Blaney, W. M. Proc. 2nd Intern. Neem Conf. 1984, pp. 163-180.
100. Saxena, K. N.; Rembold, H. Proc. 2nd Intern. Neem Conf. 1984, pp. 199-210.
101. Kubo, I., Klocke, J. A. Agric. Biol. Chem. 1982, 46, 1951-1953.
102. Haasler, C. Proc. 2nd Intern. Neem Conf. 1984, pp. 321-330.
103. Sharma, H. C.; Leuschner, K.; Sankaram, A. V. B.; Gunasekhar, D.; Marthandamurthi, M.; Bhaskariah, K.; Subramanyam, M.; Sultana, N. Proc. 2nd Intern. Neem Conf. 1984, pp. 291-320.
104. Sharma, R. N.; Nagasampagi, B. A.; Bhosale, A. S.; Kulkarni, M. M.; Tungikar, V. B. Proc. 2nd Intern. Neem Conf. 1984, pp. 115-128.
105. Adhikary, S. Proc. 1st Intern. Neem Conf. 1981, pp. 215-222.
106. Hellpap, C. Proc. 2nd Intern. Neem Conf. 1984, pp. 353-364.
107. Redfern, R. E.; Warthen, J. D., Jr.; Jacobson, M.; Stokes, J. B. J. Environ. Sci. Health 1984, A19, 477-481.
108. Meisner, J.; Ascher, K. R. S.; Aly, R. Proc. 1st Intern. Neem Conf. 1981, pp. 157-170.
109. Meisner, J.; Ascher, K. R. S.; Zur, M. Phytoparasitica 1983, 11, 51-54.

110. Gujar, G. T.; Mehrotra, K. N. Indian J. Expt. Biol. 1983, 21, 292-293.
111. Sombatsiri, K.; Tigvattanont, S. Proc. 2nd Intern. Neem Conf. 1984, pp. 95-100.
112. Joshi, B. G.; Sitaramaiah, S. Phytoparasitica 1979, 7, 199-202.
113. Warthen, J. D., Jr.; Uebel, E. C. Proc. 1st Intern. Neem Conf. 1981, pp. 137-148.
114. Saxena, R. C. Indian J. Agric. Sci. 1982, 52, 51-52.
115. Adler, V. E.; Uebel, E. C. J. Environ. Sci. Health 1984, A19, 393-403.
116. Sieber, K. P.; Rembold, H. J. Insect Physiol. 1983, 29, 523-527.
117. Rembold, H.; Sieber, K. P. Proc. 1st Intern. Neem Conf. 1981, pp. 75-80.
118. Naraynan, C. R.; Singh, R. P.; Sawaikar, D. D. Indian J. Entomol. 1980, 42, 469-472.
119. Tella, A. W. Afr. Pharmacol. Drug Res. 1976, 3, 80P.
120. Arigbabu, S. O.; Don-Pedro, S. G. Afr. J. Pharm. Pharmaceut. Sci. 1971, 114, 181-184.
121. Okpanyi, S. N.; Ezeukwu, G. C. Planta Med. 1981, 41, 34-39.
122. Thompson, E. B.; Anderson, C. C. J. Pharm. Sci. 1978, 67, 1476-1478.
123. Pillai, N. R.; Santhakumari, G. Planta Med. 1984, 50, 143-146.
124. Sinha, K. C.; Piar, S. S.; Tiwary, R. S.; Dhawan, A. K.; Bardhan, J.; Thomas, P.; Kain, A. K.; Jain, R. K. Indian J. Med. Res. 1984, 79, 131-136.
125. Sharma, V. N.; Saksena, K. P. Indian J. Med. Sci. 1959, 13, 1038-1042.
126. Sinha, K. C.; Riar, S. S.; Bardhan, J.; Thomas, P.; Kain, A. K.; Jain, R. K. Indian J. Med. Res. 1984, 80, 708-710.
127. McGregor, J., personal communication
128. Vijjan, V. K.; Tripathi, H. C.; Parihar, N. S. J. Environ. Biol. 1982, 3, 47-52.
129. Qadri, S. S. H.; Usha, G.; Jabeen, K. Intern. Pest Control 1984, 26, 18-20.
130. Pillai, N. R.; Santhakumari, G. Planta Med. 1984, 50, 146-148.
131. Sinniah, D.; Baskaran, G. Lancet 1981, 487-489.
132. Schmutterer, H., personal communication.
133. Radwanski, S. A. Proc. 1st Intern. Neem Conf. 1981, pp. 267-278.
134. Lewis, W. H.; Elvin-Lewis, M. P. F. Econ. Botany 1983, 37, 69-70.
135. Larson, R., personal communication.

RECEIVED August 19, 1985

INDEXES

Author Index

Subject Index

235

Production and indexing by Keith B. Belton
Jacket design by Pamela Lewis

Elements typeset by Hot Type Ltd., Washington, DC
Printed and bound by Maple Press Co., York, PA